# Wolf and Coyote Trapping

A. R. Harding

**Wolf and Coyote Trapping**

Copyright © 2022 Bibliotech Press
All rights reserved

The present edition is a reproduction of previous publication of this classic work. Minor typographical errors may have been corrected without note; however, for an authentic reading experience the spelling, punctuation, and capitalization have been retained from the original text.

ISBN: 978-1-63637-806-0

# TABLE OF CONTENTS

| Chapter | | |
|---|---|---|
| | Introduction | iv |
| I | The Timber Wolf | 1 |
| II | The Coyote | 6 |
| III | Killing of Stock and Game by Wolves | 11 |
| IV | Bounties | 18 |
| V | Hunting Young Wolves and Coyotes | 28 |
| VI | Hunting Wolves with Dogs | 33 |
| VII | Still Hunting Wolves and Coyotes | 39 |
| VIII | Poisoning Wolves | 46 |
| IX | Trapping Wolves | 52 |
| X | Scents and Baits | 60 |
| XI | Scent Methods | 65 |
| XII | Bait Methods for Wolves | 70 |
| XIII | Southern Bait Methods for Coyotes | 79 |
| XIV | Northern Bait Methods for Coyotes | 85 |
| XV | Blind Set Methods | 96 |
| XVI | Snow Set Methods | 100 |
| XVII | Some Rules and Things to Remember | 107 |
| XVIII | The Treacherous Grey Wolf | 111 |
| XIX | Wolf Catching | 115 |
| XX | With the Coyotes | 120 |
| XXI | Wolf Trapping an Art | 127 |

# INTRODUCTION

There are certain wild animals which when hard pressed by severe cold and hunger, will raid the farmers and ranchmen's yards, killing fowls and stock. There however, are no animals that destroy so much stock as wolves and coyotes as they largely live upon the property of farmers, settlers and ranchmen to which they add game as they can get it.

While these animals are trapped, shot, poisoned, hunted with dogs, etc., their numbers, in some states, seem to be on the increase rather than the decrease in face of the fact that heavy bounties are offered.

The fact that wolf and coyote scalps command a bounty, in many states, and in addition their pelts are valuable, makes the hunting and trapping of these animals of no little importance.

One thing that has helped to keep the members of these "howlers" so numerous is the fact that they are among the shrewdest animal in America. The day of their extermination is, no doubt, far in the distance.

This book contains much of value to those who expect to follow the business of catching wolves and coyotes. A great deal of the habits and many of the methods were written by Mr. E. Kreps, who has had experience with these animals upon the Western Plains, in Canada, and the South. Additional information has been secured from Government Bulletins and experienced "wolfers" from various parts of America.

A. R. Harding

# CHAPTER I

# THE TIMBER WOLF

Wolves of all species belong to that class of animals known as the dog family, the members of which are considered to be the most intelligent of brute animals. They are found, in one species or another, in almost every part of the world. They are strictly carnivorous and are beyond all doubt the most destructive of all wild animals.

In general appearance the wolf resembles a large dog having erect ears, elongated muzzle, long heavy fur and bushy tail. The size and color varies considerably as there are many varieties.

The wolves of North America may be divided into two distinct groups, namely, the large timber wolves, and the prairie wolves or coyotes (ki'-yote). Of the timber wolves there are a number of varieties, perhaps species, for there is considerable difference in size and color. For instance there is the small black wolf which is still found in Florida, and the large Arctic wolf which is found in far Northern Canada and Alaska, the color of which is a pure white with a black tip to the tail. Then there is that intermediate variety known as the Grey Wolf, also called "Timber Wolf," "Lobo" and "Wolf," the latter indefinite name being used throughout the West to distinguish the animal from the prairie species. It is the most common of the American wolves, the numbers of this variety being in excess of all of the others combined. In addition to those mentioned, there are others such as the Red Wolf of Texas and the Brindled Wolf of Mexico. All of these, however, belong to the group known to naturalists as the Timber Wolves. Just how many species and how many distinct varieties there are is not known.

As a rule, the largest wolves are found in the North; the Gray Wolves of the western plains being slightly smaller than the white and Dusky Wolves of Northern Canada and Alaska, specimens of which, it is said, sometimes weigh as much as one hundred and fifty pounds. Again the wolves of the southern part of the United States and of Mexico are smaller than the gray variety.

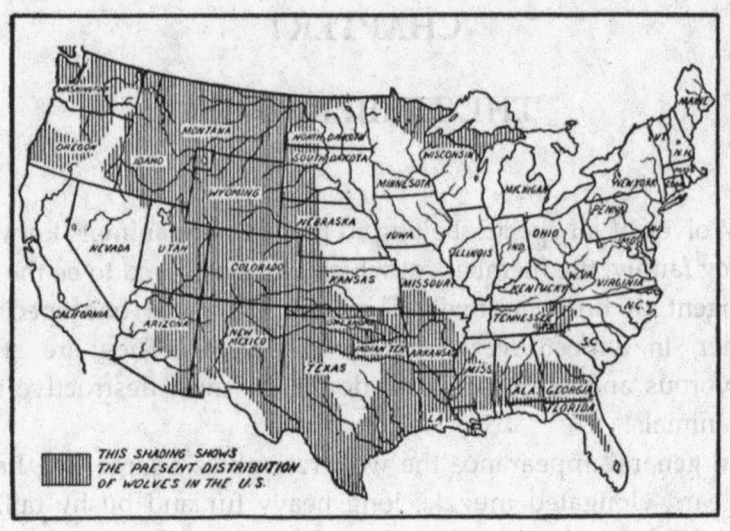

*The Range of the Timber Wolf*

The average full grown wolf will measure about five feet in length, from the end of the nose to the tip of the tail, and will weigh from eighty to one hundred pounds, but specimens have been killed which far exceeded these figures. The prevailing color is gray, being darkest on the back and dusky on the shoulders and hips. The tail is very bushy and the fur of the body is long and shaggy. The ears are erect and pointed, the muzzle long and heavy, the eyes brown and considering the fierce, bloodthirsty nature of the animal, have a very gentle expression.

In early days wolves were found in all parts of the country but they have been exterminated or driven out of the thickly settled portions and their present distribution in the United States is shown by the accompanying map. As will be noted they are found in only a small portion of Nevada and none are found in California, but they are to be met with in all other states west of the Missouri and the lower Mississippi, also all of the most southern tier of states, as well as those parts bordering on Lake Superior. A few are yet found in the Smokey Mountains of North Carolina and Tennessee. They are probably most abundant in Northern Michigan and Northern Minnesota, Western Wyoming, Montana and New Mexico.

Wyoming is the center of the wolf infested country and they are found in greatest numbers in that state, on the headwaters of the Green River. As to the numbers still found the report of the Biological Survey for the years 1895 to 1906, inclusive, but not including the year 1898, shows that bounties were paid on 20,819 wolves in that state.

In Northern Michigan they are also abundant. In the year 1907, thirty-four wolves were killed in Ontonagon County; in Luce County fifty-four were killed up to November 10th, '07, and in Schoolcraft Co., thirty were killed from October 1st, '07 to April 29th, '08. This gives a total of one hundred and eighteen wolves killed in three out of the sixteen counties of the Upper Peninsula. These statistics are from a pamphlet issued by the Department of Agriculture.

The breeding season of the timber wolves is not as definite as that of many of the furbearing animals, for the young make their appearance from early in March until in May, and an occasional litter will be born during the summer, even as late as August. The mating season of course varies, but is mainly in January and February, the period of gestation being nine weeks. The number in a litter varies from five to thirteen, the usual number being eight or ten.

In early days the wolves of the western plains followed the great buffalo herds and preyed on the young animals, also the old and feeble. After the extermination of that animal they turned their attention to the herds of cattle which soon covered the great western range and their depredations have become a positive nuisance. In the Northern States and throughout Canada they subsist almost entirely on wild game.

Wolves den in the ground or rocks in natural dens if such can be found, but in case natural excavations are rare as in northern portions of the country, they appropriate and enlarge the homes of other animals. In the heavily timbered country they sometimes den in hollow logs.

The wolf is both cowardly and courageous, depending on circumstances. When found singly, and especially in daylight the animal is as much of a coward as any creature could possibly be, and especially does it fear man. But when suffering from the pangs of hunger and when traveling in bands as they usually do, they are bold, fierce and bloodthirsty creatures. In such cases they have been known to attack man.

When hunting large game, wolves always go in bands, usually of three to five but often a larger number. They invariably kill animals by springing on from behind and hamstringing the victim. Small game is hunted by lone animals.

The great losses suffered by stockmen in the West led the Biological Survey, in connection with the Forest Service of the Department of Agriculture, to make a special investigation, and later a general campaign against the wolves of the National Forests began. During the year 1907 a large number of wolves and coyotes were captured in and near the forest reserves: the number from the various states being as follows:

| STATE | WOLVES | COYOTES |
|---|---|---|
| Wyoming | 1,009 | 1,983 |
| Montana | 261 | 2.629 |
| Idaho | 14 | 3,881 |
| Washington | 10 | 675 |
| Colorado | 65 | 2,362 |
| Oklahoma | 3 | 15 |
| New Mexico | 232 | 544 |
| Arizona | 127 | 1,424 |
| Utah |  | 5,001 |
| Nevada |  | 500 |
| California |  | 224 |
| Oregon | 2 | 3,290 |
| Total | 1,723 | 22,528 |

Many of these animals were captured by the forest guards but in addition the government employed a number of expert trappers.

On the Gila National Forest 36 wolves and 30 coyotes were killed by one forest guard, who sent the skulls to the Biological Survey for identification, as well as the skulls of 9 bears, 7 mountain lions, 17 bobcats, and 46 grey foxes. One den of 8 very young wolf pups was taken March 13. These statistics are from Circular 63, issued by the U. S. Department of Agriculture.

Wolves are great ramblers, traveling over a large section of country. Like almost all other animals of rambling habits, they have their regular routes of travel. By this, we mean they follow the same valleys, passes, water courses, etc., but when in pursuit of game they sometimes stray quite a long distance out of their course.

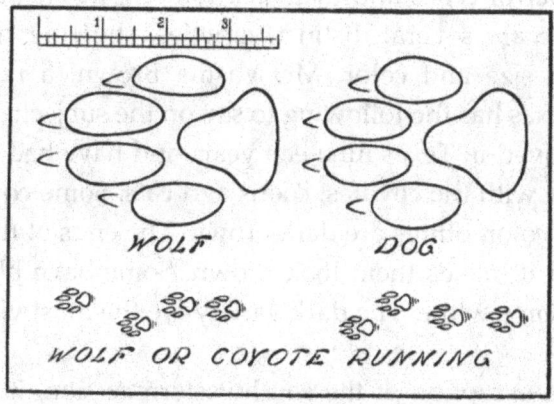

Track of the Grey Wolf, Compared With That of a Dog

The track of the wolf resembles that of a dog, but is a trifle narrower in proportion to its length. The difference is in the two middle toes, which are somewhat longer on the wolf, however, the difference is so slight that it could easily pass unnoticed. When the wolf is running these toes are spread well apart. The length of step when the animal is walking will be from 18 to 24 inches, and the average footprint will measure about 2 3/4 or 3 inches in width by about 3 1/2 or 4 inches in length. Ernest Thompson Seton, the naturalist claims that he can judge with fair accuracy, the weight of a wolf by the size of the track. He allows twenty pounds for each inch in length, of the foot print.

# CHAPTER II

# THE COYOTE

In the western parts of the United States, the coyote is far more abundant than the grey, or timber wolf, but its range is more limited as it is found only in those parts lying west of the Mississippi River and in the western portion of the Dominion of Canada. As there are a number of varieties of the timber wolf, so it is with the coyote, but naturalists have never yet been able to agree on the number of types and their distribution. In the Southwest, it appears there are several distinct varieties, showing considerable difference in size and color. Mr. Vasma Brown, a noted coyote trapper of Texas has the following to say on the subject:

"I have lived in Texas nineteen years and have had some years of experience with the coyotes, coons and cats. Some coyotes are of a silver-grey color, others are dark brown. The ends of their hair are jet black and it makes them look brown. Some have black tips on the tail and some white. The dark variety are the most vicious of the two."

With the exception of the southwestern section, it is probable that the coyotes of all portions of the Great Plains and the country to the westward are of the same variety, and a description of this, the most common type will answer for the species. In size, the coyote or prairie wolf is considerably smaller than the timber wolf, the largest specimens of the former being about equal in size to the smallest adult wolves. The average coyote will measure about thirty-six or thirty-eight inches from the end of the nose to the base of the tail, which is about sixteen inches additional length. The fur is of about the same texture as that of the grey fox and the general color is fulvous, black and white hairs being mingled in parts, giving a grizzled appearance. The ears are larger, comparatively than those of the grey wolf, and the muzzle is more pointed. All through the animal appears to be of more delicate build. A larger form of the coyote is found in Minnesota and the adjoining territory and is commonly known as the "brush wolf". Whether this is a distinct variety is not known.

Coyotes are intelligent and cunning animals and their habits and general appearance suggest the fox rather than the wolf. While they are greedy, bloodthirsty creatures, they are sneaking and cowardly and never kill animals larger than deer, in fact they rarely attack such large game. An Arizona trapper writes:

"The coyote bears the same relation to the wolf family that the Apache Indian does to the human race. It is a belief among some of the Apaches that they turn into coyotes when they depart this life, and nothing will induce one of them to kill a coyote. Like the Indian he is sneaky and treacherous, and full of the devil."

While there is no doubt that the animal enjoys its wild, free life, it always has a miserable, distressed expression. It carries its tail in a drooping manner and slinks out of sight like a dog that has been doing wrong and has a troubled conscience.

The high piercing cry of the animal, which is so different from the deep bass note of the timber wolf, is mournful in the extreme. In the morning before the coyotes retire for the day, they stop on the top of some elevation and sound their "reveille", which once heard will never be forgotten. It is a shrill, piercing note, combining a howl with a bark and although in all probability there will be only a pair of the animals, one who does not know would be inclined to think that the number is larger, the notes are so commingled.

Coyotes live in natural dens in the rocks, also in dens of badgers, in the prairie country. In the "Bad Lands" of the West and the foot hills of the mountain ranges, wind worn holes in the rimrock and buttes are quite common and the animals have no trouble in securing a good den. Naturally, they select the most secluded and inaccessible places for their dens. The food of the coyote consists of small game, such as hares and grouse, prairie dogs and any other small animals that they can capture. In the sheep raising districts of the Western States they are very destructive to sheep and in those parts it is probable that their food consists mostly of mutton. They feed on carrion and have a particular liking for horse flesh. They also kill badgers and when conditions are very favorable may kill an occasional deer or antelope. They also sometimes kill calves and hogs.

Speaking of conditions in Oregon and other parts of the Northwest, one of our friends writes:

"The prairie wolf or coyote in the Western states are becoming so numerous that it looks as though the sheep industry in Idaho and Eastern Oregon would soon be a thing of the past, if something it not done to lesson the number of the destructive coyotes.

"Twenty years ago there were a great many coyotes in Oregon, but the black tail rabbits were so numerous then that the coyote contented himself with them and did not molest the sheep to any great extent. Idaho and Oregon both put a bounty on rabbits, which soon caused them to become scarce, then the coyotes began their depredations among the sheep. The wool growers supplied themselves with plenty of strychnine and kept the coyote reduced to quite an extent. Of late years it seems that poison will not kill a coyote. As soon as he feels the effect of the poison he throws up the bait he has just eaten, and in a few minutes he is all right.

The only way to kill coyotes these days is with the gun, the trap or with dogs. They are so thick here now that hounds would not be much good, as the coyotes would change at any time and run them down. I don't think there was a band of sheep anywhere in this country but what suffered more or less from coyotes last winter. I trapped some last winter for the Munz Brothers, and I saw where 48 sheep had been killed at one camp. They had been camped there about ten days. This is about an average killing if the weather is stormy.

"In Southeastern Oregon there is a desert about one hundred miles square, and thirty or forty bands of sheep feed there every winter. They run from two to three thousand sheep in a band. The sheep men on this desert last winter, 1904-'05, paid $40.00 per month and board for trappers to trap coyotes, and the trappers were allowed to keep the furs they caught. Some of them made very large wages."

It is said that when hunting rabbits, two coyotes will join forces and in this way one animal will drive the game to within reach of the other, thus avoiding the fatigue caused by running down game. Naturalists also claim that the adult animals will sometimes drive the game close to the den, so that the young coyotes may have the opportunity of killing it. They frequently pick up scraps about the camps, and if undisturbed, will in a short time, lose much of their timidity. Old camping places are always inspected in the hopes of finding some morsel of food, and one can always find coyote tracks in the ashes of the campfire.

Though the coyote belongs to the flesh-eating class of animals, it is not strictly carnivorous. In late summer when the wild rose tips are red and sweet and berries are plentiful, its flesh eating propensities forsake it in part and it adds fruit to its "bill of fare". Whether this is caused by hunger or a change of appetite, or whether the fruit acts as a tonic and the animal, instinctively, realizes that it must tone up its system in preparation for the long winter, is not known.

Coyotes have a more regular breeding season than the timber wolves, for practically all of the young make their appearance in the months of April and May. The number of young varies from five to twelve. The young animals are of a yellowish grey color with brown ears and black tail, muzzle tawny or yellowish brown. As they become older they take on a lighter shade and the tail changes to greyish with a black tip. Both wolves and coyotes pair for the breeding season and the males stay with the females during the summer and help take care of the young. It is probable that they do not breed until two years of age. As soon as the young are strong enough, and their eyes are open they commence to play about the mouth of the den and later on the mother leads them to the nearest water and finally allows them to accompany her on hunting excursions. In late summer they start out to shift for themselves.

As before mentioned, the coyote is a wary and cunning animal, especially in the more settled portions of its range; where man is not too much in evidence, they are far less wary. Again the fact that there are several varieties may account for the difference in the nature of the animals of the various sections, anyway those of the southern part of the range are less wary than those of the North.

The trappers of Texas, Arizona and New Mexico claim that the coyote is a fool and is easily caught while those of the North and Northwest find them exceedingly cunning and intelligent. Not only does the animal appear to know when you are armed but it also seems to know something of the range of the weapon and will sneak along provokingly close, but just out of reach. When one is unarmed they appear to be more bold and will loaf around in the most unconcerned manner imaginable.

In intelligence and cunning, we consider the northern coyote the equal of the eastern red fox. While the western trappers make very large catches of coyotes, we believe that if foxes were found in equal numbers the catches of those animals would be fully as large. The number of coyotes found in some parts of the West is almost incredible, and in most parts one will find a hundred coyotes to one grey wolf.

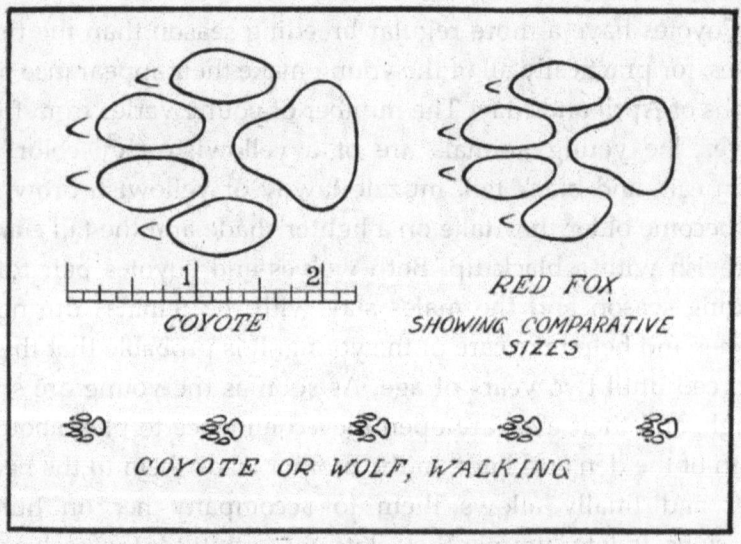

*Track of the Coyote*

The coyote makes a track similar to that of the timber wolf, but considerably smaller. The length of step, when walking, is about sixteen inches and the footprints will measure about two or two and a fourth inches in length by one and a half in width.

# CHAPTER III

# KILLING OF STOCK AND GAME BY WOLVES

Undoubtedly the wolves and coyotes of the United States and Canada destroy more stock and game than all other predatory animals combined. In the Western part of our country where stock raising is one of the principal industries, the ranchmen suffer great losses from the depredation of these animals, and in other sections the wolves destroy large quantities of game. The reason that wolves are more destructive than others of the carnivora is that when they have the opportunity, they kill far more than they can consume for food. Often they only tear a mouthful of flesh from the body of their victim; sometimes they do not even kill the animal but leave it to suffer a slow and painful death. The animals that are only slightly bitten are sure to die from blood poisoning, according to the western ranchmen.

*Wolves Killing a Deer*

The wolf's method of attack is from the rear, springing on its victim and hamstringing it and literally eating it alive. The bite of the wolf is a succession of quick, savage snaps and there is no salvation for the creature that has no means of defense from a rear attack. This peculiar method of killing prey can not be practiced successfully on horses, owing to the fact that they can defend themselves by kicking, but for all of that, a considerable number of colts and a few full grown horses are killed. For this reason cattle suffer more than horses, but while the horse is, to a certain extent, exempt from attack by wolves, they are frequently killed by mountain lions, because their method of attack, a spring at the head and throat is more successful with these animals than with cattle. As food, however, horse flesh is preferred to beef by both of these animals.

One of the western trappers writes:

"Many times in the past thirty years I have watched wolves catch cows. The wolf is by nature a coward and will not, singly, attack a grown cow, though he will by himself kill a pig, chicken, calf, goat or sheep.

"On the ranges, where the stockmen and settlements are far apart, wolves go in bunches, from three to ten or even more, and when very hungry a bunch of them will attack a grown bull. They frighten him by snapping and playing around him till they get him on the run, when the bunch give full chase and stay close at his heels. While he is running in this way, one or more of them will grab him by the ham strings just above the hock joint. The bull makes, of course, a vigorous effort to free himself from the wolf, but before he can do so, the sharp teeth of the latter have cut or partially cut the ham string. They keep him on the run till they finally cut him down in both ham strings and then he cannot go further or fight the hungry wolves off.

"The whole bunch then eat his hams out while the bull is still alive, and after they get their full they let him rest. When they want to fill up again, they return and eat him till he dies, finishing the carcass as they require food.

"I have seen horses and cattle killed by wolves in this way live for several days with their hams eaten out, and have never seen the wolf make his attack or give chase in any other way. Being cowardly, he always follows behind and keeps out of all danger from the bull's hoofs."

Of cattle, calves and yearlings are generally selected, partly because the flesh of the younger animals is more to the wolf's liking and partly because they cannot defend themselves as readily as full grown animals, but full grown steers are also killed at times. Far more cattle are killed than are eaten. The wolf prefers fresh food always and in summer when their resources are unlimited they seldom return to the carcass for a second meal.

In "Bulletin 72," issued by the United States Department of Agriculture, the author, Mr. Vernon Bailey, has the following to say on the subject:

"The actual number of cattle killed by wolves can not be determined. Comparatively few animals are found by cattlemen and hunters, when freshly killed, with wolf tracks around them and with wolf marks on them. Not all of the adult cattle missing from a herd can surely be charged the depredations of wolves, while missing calves may have been taken by wolves, by mountain lions, or by 'rustlers.'"

Nevertheless there are data enough from which to draw fairly reliable conclusions. In the Green River Basin, Wyoming, on April 2, 1906, Mr. Charles Budd had 8 yearling calves and 4 colts killed in his pasture by wolves within six weeks. At Big Piney a number of cattle and a few horses had been killed around the settlement during the previous fall and winter. At Pinedale, members of the local stockmen's association counted 30 head of cattle killed in the valley around Cora and Pinedale in 1905, between April, when the cattle were turned out on the range, and June 30, when they were driven to the mountains. In 1906, wolves were said to have come into the pastures near Cora and Pinedale and begun killing cattle in January on the "feed grounds," and Mr. George Glover counted up 22 head of cattle killed by them up to April 10. Just north of Cora, Mr. Alexander, a well known ranchman, told me that the wolves killed near his place in June, 1904, a large three year old steer, a cow, 3 yearlings and a horse. On the G O S Ranch, in the Gila Forest Reserve in New Mexico, May 11 to 30, 1906, the cowboys on the round-up reported finding calves or yearlings killed by wolves almost daily, and Mr. Victor Culberson, president of the company, estimated the loss by wolves on the ranch at 10 per cent. of the cattle.

In a letter to the Biological Survey, under date of April 3, 1896, Mr. R. M. Allen, general manager of the Standard Cattle Company, with headquarters at Ames, Neb., and ranches in both Wyoming and Montana, states that in 1894 his company paid a $5.00 bounty at their Wyoming ranch on almost exactly 500 wolves. The total loss to Wyoming through the depredations of wolves Mr. Allen estimated at a million dollars a year.

In an address before the National Live Stock Association at Denver, Col., January 25, 1899, Mr. A. J. Bothwell said: "In central Wyoming my experience has been that these wolves kill from 10 to 20 per cent. of the annual increase of the herds."

Lieut. E. L. Munson, of Chouteau County, Mont., writing in Recreation, says: "It is said that in this country the loss from wolves and coyotes is about 15 per cent. * * * Wolves in this vicinity seldom kill sheep, as the latter are too carefully herded. They get a good many young colts, but prey especially on young cattle."

Mr. J. B. Jennett, of Stanford, Montana, says in Recreation: "A family of wolves will destroy about $3,000 worth of stock per annum."

The loss caused by wolves and coyotes in Big Horn County, Wyo., is estimated at three hundred thousand dollars per year. It has been variously estimated that each grey wolf costs the stockmen from two hundred and fifty to one thousand dollars annually.

Sheep, for some reason, are seldom troubled by timber wolves in the West, but suffer considerably from the attacks of coyotes; in fact, the loss occasioned the sheep men of Wyoming and Montana in this way is enormous. In summer when the sheep are driven up into the mountains, the coyotes migrate to those sections and kill sheep whenever the opportunity is presented. In the fall when the sheep are brought down into the foothills, the coyotes are also to be found in great numbers in those parts. In all probability there is a greater loss occasioned by the depredations of coyotes in the two states mentioned than is caused by wolves and mountain lions combined. Farther south, however, it is the wolf that does the most mischief. Where timber wolves are plentiful and very little stock is raised, as in the northern parts of Michigan, Wisconsin and Minnesota, sheep are not safe from the attacks of wolves, and for that reason few sheep are raised in those parts. It is probably the fact that the western range is very open and the sheep always carefully guarded by herders that they suffer so little from timber wolves in the Western States.

In the swamps of the Southern States, and especially in the lowlands of Texas, Louisiana and Arkansas, hogs are sometimes killed by wolves. In New Mexico, Texas, Colorado and Mexico where large numbers of goats are raised, these animals are frequently killed.

That an immense amount of game is killed in the wilder and less thickly settled portions of the United States and Canada goes without saying. In the West the wild game does not suffer as much as does the domestic animals, but in the heavily timbered portions of the country where little stock is raised as in the states bordering on Lake Superior and in the greater part of Canada large numbers of deer and other game animals fall victims to these fierce creatures. Regarding the killing of game on the western cattle range, Mr. Bailey has the following to say:

"At Big Piney, Wyoming, I examined wolf dung in probably fifty places around dens and along wolf trails. In about nine-tenths of the cases it was composed mainly or entirely of cattle or horse hair; in all other cases but one, of rabbit fur and bones, and in this one case mainly of antelope hair. A herd of 20 or 30 antelope wintered about 5 or 6 miles from this den, and the old wolves frequently visited the herd, but I could find no other evidence that they destroyed antelope, though I followed wolf tracks for many miles among the antelope tracks on the snow. Jack rabbits were killed and eaten along the trails or brought to the den and eaten near it almost every night, and a half eaten cottontail was found in the den with the little pups. While wolves are usually found around antelope herds, they are probably able to kill only the sick, crippled and young. The following note from Wyoming appeared in the Pinedale Roundup of July 4, 1906:

While riding on the outside circle with the late round-up, Nelse Jorgensen chanced to see a wolf making away with a fawn antelope. He gave chase to the animal, but it succeeded in getting away, never letting loose on its catch.

About a den near Cora, the numerous deposits of wolf dung on the crest of the ridge not far away were found to be composed of horse and cattle hair, though fresh elk tracks were abundant over the side hills on all sides of the den, while cattle and horses were then to be found only in the valley, 8 miles distant. Several jack rabbits had been brought in and eaten and the old wolf on her way to the den had laid down her load, evidently a jack rabbit, gone aside some 20 feet and caught a ruffed grouse eaten it on the spot, and then resumed her load and her journey to the waiting pups. One small carpal bone in this den may have been from a deer or small elk, but no other trace of game was found.

Talking with hunters and trappers who spend much time in the mountains when the snow is on the ground brought little positive information on the destruction of elk or deer by wolves. Mr. George Glover, a forest ranger long familiar with the Wind River Mountains in both winter and summer, said that he had found a large blacktail buck which the wolves had eaten, but he suspected that it had been previously shot by hunters. In many winters of trapping where elk were abundant, Mr. Glover has never found any evidence that elk had been killed by wolves. Coyotes constantly follow the elk herds, especially in spring when the calves are being born, and probably destroy many of the young, but wolves apparently do not share this habit. It seems probably, however, that in summer the young of both elk and deer suffer to some extent while the wolves are among them in the mountains."

In the Northern Peninsula of Michigan, wolves are very plentiful and large numbers of deer are killed during the winter months, the remains being found later by hunters, trappers, and lumbermen. The same conditions exist in Northern Minnesota and Wisconsin, also in parts of Ontario, Canada. In the Rainy River District, wolves have always been abundant and much game has been killed by them. Farther east, they are just making their appearance of late years having followed the deer which are coming into the country from some other section. Farther east, in the eastern portions of New Ontario and in some parts of Quebec wolves are also numerous. One of our friends from Northern Wisconsin writes as follows:

"I have trapped and caught five old female wolves since I came to Iron County, Wisconsin, six years ago. Two of them I got in Michigan, Gogebic County, as I live almost on the line. There are times when you can see six or eight wolf tracks all going down the river or coming up at the same time. You can go again for a week and never see a track. I have followed them for a week, in deep snow on snow shoes, and never left their track, and in one week I set traps at 50 different deer that wolves had killed. I might have gotten a few more wolves but the fox, mink, cats, skunk, owls and "porkys" (porcupines) were bound to butt in. At one set I got a wolf, 3 foxes, 1 skunk, 1 mink and 10 porkys till June.

Two wolves caught a buck that would weigh 150 pounds, within 10 rods of my camp one night. The next morning there was not one pound of meat left on the bones.

I had a tent and one shanty in Gogebic County last winter, and I know the wolves killed 500 deer on the snow. How many fawns and does did they kill in summer time when you cannot see their tracks? The wild cats are not so bad, a fawn, rabbit or partridge makes a meal for them."

In the far north where the barren ground caribou is the principal game animal, and where wolves are plentiful, there can be no doubt that they kill large numbers of those animals. Musk oxen are also killed, and farther south the moose is killed by wolves, but it is our belief that the number is comparatively small. The moose is such a large and powerful animal that even a band of half starved wolves will, as a rule, pass it by, but there can be no doubt of the fact that they do kill them on rare occasions.

The elk is a great enemy of the wolf and it appears that they are seldom molested. Beyond all doubt the deer is the principal prey of the timber wolf.

# CHAPTER IV

# BOUNTIES

For many years the state governments of the wolf infested country have been paying bounties on wolves and coyotes, to encourage the hunting and trapping of these animals. It is doubtful, however, whether the bounties offered are sufficient to encourage any, other than the regular trappers, to hunt wolves, and if they are, it has certainly had no definite results, for the wolves and coyotes, taken over the whole country, are practically as plentiful as ever.

Realizing that the state bounties were not a sufficient inducement to trappers, certain of the counties of those states where wolves are most abundant, offer additional bounty. This has the effect of thinning the wolves out of that county alone, but they immediately become more plentiful in the adjoining portions of the country.

In some of the Western States, the stockmen pay a bounty, in addition to that offered by the state. Some of them even offer special inducements, in addition to the bounties paid on the captured animals, and among them may be mentioned, board and lodging for the trapper, bait for the traps and the use of saddle and pack horses.

Such special offers to trappers have the effect of stimulating the hunting and trapping of noxious animals in that immediate vicinity and the result is, a thinning out of the animals for the time being. Usually the trappers drift into those sections where the animals are most plentiful and the bounty is highest.

One of the Government bulletins has the following to say regarding the bounty question:

"Bounties, even when excessively high, have proved ineffective in keeping down the wolves, and the more intelligent ranchmen are questioning whether the bounty system pays. In the past ten years Wyoming has paid out in State bounties over $65,000 on wolves alone, and $160,156 on wolves, coyotes and mountain lions together, and to this must be added still larger sums in local and county bounties on the same animals."

"In many cases three bounties are paid on each wolf. In the upper Green River Valley the local stockmen's association pays a bounty of $10 on each wolf pup, $20 on each grown dog wolf, and $40 on each bitch with pup. Fremont County adds $3 to each of these, and the State of Wyoming $3 more. Many of the large ranchers pay a private bounty of $10 to $20 in addition to the county and state bounty. Gov. Bryant B. Brooks, of Wyoming, paid six years ago, on his ranch in Natrona County, $10 each on 50 wolves in one year, and considered it a good investment, since it practically cleared his range of wolves for the time. It invariably happens, however, that when cleared out of one section the wolves are left undisturbed to breed in neighboring sections, and the depleted country is soon restocked."

"A floating class of hunters and trappers receive most of the bounty money and drift to the sections where the bounty is highest. If extermination is left to these men, it will be a long process. Even some of the small ranch owners support themselves in part from the wolf harvest, and it is not uncommon to hear men boast that they know the location of dens, but are leaving the young to grow up for higher bounty. The frauds, which have frequently wasted the funds appropriated for the destruction of noxious animals almost vitiate the wolf records of some of the States: If bounties resulted in the extermination of the wolves or in an important reduction in their number, the bounty system should be encouraged, but if it merely begets fraud and yields a perpetual harvest for the support of a floating class of citizens, other means should be adopted."

The failure of bounties to accomplish their proposed object was clearly shown by Dr. T. S. Palmer in 1896. Under the heading, "What have bounties accomplished," he says:

"Advocates of the bounty system seem to think that almost any species can be exterminated in a short time if the premiums are only high enough. Extermination, however, is not a question of months, but of years, and it is a mistake to suppose that it can be accomplished rapidly except under extraordinary circumstances, as in the case of the buffalo and the fur seal. Theoretically, a bounty should be high enough to insure the destruction of at least a majority of the individuals during the first season, but it has already been shown that scarcely a single State has been able to maintain a high rate for more than a few months, and it is evident that the higher the rate, the greater the danger of fraud. Although Virginia has encouraged the killing of wolves almost from the first settlement of the colony, and has sometimes paid as high as $25 apiece for their scalps, wolves were not exterminated until about the middle of this (the past) century, or until the rewards had been in force for more than two hundred years. Nor did they become extinct in England until the beginning of the sixteenth century, although efforts toward their extermination had been begun in the reign of King Edgar (959-975). France, which has maintained bounties on these animals for more than a century, found it necessary to increase the rewards to $30 and $40 in 1882, and in twelve years expended no less than $115,000 for nearly 8,000 wolves."

"The larger animals are gradually becoming rare, particularly in the East, but it can not be said that bounties have brought about the extermination of a single species in any State."

"New Hampshire has been paying for bears about as long as Maine, but in 1894 the State treasurer called attention to the large number reported by four or five of the towns, and added that should the other 231 towns 'be equally successful in breeding wild animals for the State market, in proportion to their tax levy, it would require a State tax levy of nearly $2,000,000 to pay the bounty claims' Even New York withdrew the rewards on bears in 1895, not because they had become unnecessary, but because the number of animals killed increased steadily each year."

"Wolf skins are often ruined by the requirements of bounty laws, especially when the head, feet, or ears are cut off. The importance of preserving the skins in condition to bring the highest market price is as great as that of making it impossible to collect bounties twice. A slit in the skin can be sewed up so that it will never show on the fur side, but can not be concealed on the inside.

A single longitudinal or vertical slit, or double or cross slits 4 inches long, in the center where the fur is longest, would serve every purpose of the law without seriously impairing the market value of the skin."

One thing that is detrimental to the success of the bounty system, is the invariable "red tape" connected with such laws. In some states the bounty regulations are so complicated and so exacting, that trappers do not care to follow "wolfing" because of the trouble in securing the bounty money.

It would be impossible, in a work of this kind, to give the bounty laws of the different states, also as they are repealed so frequently, detailed information on that subject would be of little value to the prospective hunter or trapper. We give, however, an outline of the regulations in some of the principal wolf states.

The State of Wyoming pays a bounty of five dollars each on timber wolves and mountain lions, and one dollar and twenty-five cents for each coyote. In addition to this, there are both county and stockmen's bounties in certain parts of the state. Some ranchmen offer as much as forty-five dollars each, for grey wolves caught on their ranches.

In order to secure the state bounty, one must present the entire skin to the County Clerk, or Notary Public, of the county in which the animal was killed, and accompanied by affidavit to the effect that the animal was killed in that county, by the person presenting the skin, on or after March 1st, 1909. The skin must have the feet and upper jaw or head, with both upper and lower lips attached. The head will then be cut off and destroyed by the county official. Applicants for bounty must be identified.

With regard to private bounties, one should consult the county officials, but these, and in that case, the state bounty also, are as a rule, paid by the treasurer of the association offering the bounty.

Wisconsin pays twenty dollars on old wolves and eight dollars each on pups. Half of this bounty money is paid by the state and the other half by the county. In order to secure it, the trapper must take the carcass of the animal to the Town Chairman and remove the scalp in his presence. He gives a certificate to that effect and the bounty claimant presents the scalp and certificate to the County Clerk, who destroys the scalp and gives an order to the County Treasurer for one-half of the bounty. The County Clerk also sends an affidavit to the State Treasurer, stating that you have presented the scalp and it has been destroyed and the claimant then receives the balance of the bounty money from the state.

In the State of Washington the bounty is fifteen dollars on timber wolves and one dollar on coyotes. The method of procuring the bounty as given here is copied direct from the game law pamphlet:

"Upon the production to the county auditor of any county of the entire hide or pelt and right fore leg to the knee joint intact of any cougar, lynx, wild cat, coyote or timber wolf, killed in such county, each of which hides or pelts shall show two ears, eye holes, skin to tip of nose, and right fore leg to the knee join intact, the county auditor shall require satisfactory proof that such animal was killed in such county. When the county auditor is satisfied that such animal was killed in his county, he shall cut from such hide or pelt the bone of the right fore leg to the knee as aforesaid which shall be burned in the presence of such auditor and one other county official, who shall certify to the date and place of such burning."

Utah pays a bounty of ten dollars on grey wolves and two dollars and fifty cents on coyotes. The entire skin, with tail, feet and the bones of the leg, to the knee, must be presented to the County Clerk within sixty days of the date on which said animal was killed. The County Clerk must then remove and destroy the bones of the legs and the applicant will sign an affidavit stating that the animal was killed by himself, in that county and within sixty days prior to that date.

The county official will then send a certified statement to the State Auditor, along with the other papers, who, after same have been examined, will transmit the bounty money to the claimant.

No bounty will be paid on the skin of a grey wolf until it has been seen and passed upon by the board of county commissioners at their first regular meeting. Bounty claimants must be identified by a reputable citizen and tax payer of the county.

In Minnesota the bounty on grown wolves is seven dollars and fifty cents and one dollar for wolf pups. The bounty regulations are practically the same as in the other states; the entire skin with head and ears intact must be presented to the Town Clerk within thirty days and the applicant must take affidavit as to the date and place of the killing.

In other states, if our information is correct, the bounties at present (1909) are as follows:

|  | ADULT WOLVES | YOUNG WOLVES | COYOTES |
|---|---|---|---|
| Arizona | $10 00 |  | $2 00 |
| Arkansas | 5 00 |  |  |
| Colorado | 5 00 |  | 1 00 |
| Idaho | (?) 10 00 |  | (?) 1 00 |
| Kansas | 5 00 |  | 1 00 |
| Michigan | 25 00 | $10 00 |  |
| Montana | 10 00 |  | 3 00 |
| Nebraska | 4 00 |  | 1 25 |
| New Mexico | 20 00 |  | 2 00 |
| North Dakota | 4 50 |  | 2 50 |
| Oregon | 10 00 |  | 7 00 |
| South Dakota | 5 00 |  | 1 50 |
| THE CANADIAN PROVINCES |  |  |  |
| Alberta | 10 00 | 1 00 | 1 00 |
| British Columbia | 15 00 |  | 2.00 |
| Ontario | 15 00 |  |  |
| Quebec | 15 00 |  |  |
| Saskatchewan | 3 00 |  | 1 00 |

The fraud so often practiced by unscrupulous parties has always been detrimental to the efficacy of the bounty system. The Bureau of Biological Survey, have issued a special circular on this subject and being of general interest, it is reprinted here.

## WASTE IN BOUNTY PAYMENTS

"The bounty system has everywhere proved an incentive to fraud, and thousands of dollars are wasted annually in paying bounties on coyote scalps offered in place of wolves, and on the scalps of dogs, foxes, coons, badgers, and even cats, which are palmed off for wolves and coyotes. If in all states having the bounty system whole skins, including nose, ears, feet, and tail of both adult and young animals, were required as valid evidence for bounty payments, the possibility of deception would be reduced to a minimum. The common practice of paying bounty on scalps alone, or in some cases merely the ears, is dangerous, as even an expert can not always positively identify such fragments. A satisfactory way of marking skins on which the bounty has been paid is by a slit 4 to 6 inches long between the ears. This does not injure the skins for subsequent use. If all bounty-paying states would adopt such a system, the possibility of collecting more than one bounty on the same skin in different states would be avoided."

"The following directions have been prepared as an aid to county and state officers in identifying scalps, skins, and skulls of wolves and coyotes, the pups of wolves, coyotes, red, grey, and kit foxes, and young bob-cats, coons and badgers."

"The variation in dogs is so great that no one set of characters will always distinguish them from wolves or coyotes, but when there is reason to suspect that dogs are being presented for bounties, their skins and skulls should be sent to the Biological Survey for positive identification. It goes without saying that anyone detected in such fraud should be prosecuted with a view to the suppression of these dishonest practices."

## KEY TO ADULT WOLVES AND COYOTES

|  | WOLF | COYOTE |
|---|---|---|
| Width of nose pad | 1 1/4 to 1 1/2 inches | 3/4 to 1 inch |
| Width of heel pad of front foot | 1 1/2 to 2 inches | 1 inch |

Upper canine tooth—          5/10 to 6/10 inch          3/10 to 4/10
greatest diameter at base                         inch

These characters will not always hold in Oklahoma and Texas east and south of the Staked Plains, where there is a small wolf in size between the Coyote and Lobo or Plains wolf.

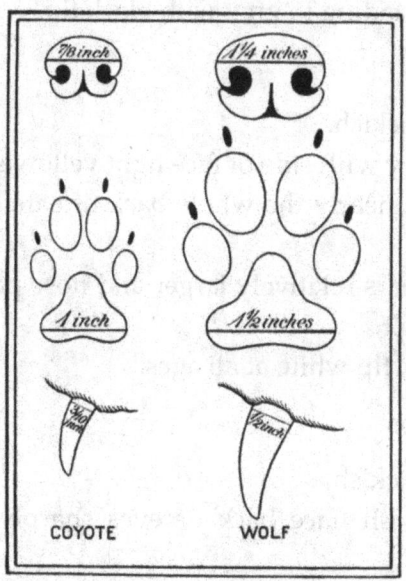

*Difference in Sizes of Noes, Heel Pads and Canine Teeth of Wolves and Coyotes*

## KEY TO WOLF, COYOTE AND FOX PUPS

### Wolf Pups

- Muzzle blackish at birth, fading in a month or 6 weeks to greyish.
- Head greyish in decided contrast to black of back, nose and ears.
- Ears black at tips, fading to greyish in a month or 6 weeks.
- Tail black, fading to grey with black tip.

### Coyote Pups

- Muzzle tawny, or yellowish brown, becoming more yellowish with age.
- Head yellowish grey, not strongly contrasted with rest of body.
- Ears dark brown at tips and back, soon fading to yellowish brown.
- Tail black, fading to grey with black tip.

### Red Fox Pups

- Muzzle blackish.
- Head dusky with side of face light yellowish.
- Ears large, nearly the whole back of ears bright black at all ages.
- Eyes and ears relatively larger and nose pad smaller than in coyote or wolf.
- Tail dusky, tip white at all ages.

### Grey Fox Pups

- Muzzle blackish.
- Head greyish, face back or eyes sharply pepper and salt grey.
- Ears large, back of ears dusky at tip, fulvous at base.
- Eyes and nose pad small.
- Tail with tip black at all ages.

### Kit Fox Pups

- Muzzle with blackish patch on each side.
- Head and face tawny or yellowish brown.
- Ears tawny without black backs or tips.
- Eyes larger and nose pad smaller than in young coyote.
- Tail with tip black at all ages.

## KEY TO YOUNG CATS, COONS AND BADGERS

- Young bobcats are much striped and spotted. Young cats of

any kind can be distinguished by the short nose and round head.
- Young coons have a broad black band across the face and eyes bordered above by a light band.
- Young badgers have a white stripe between the eyes.

The bounty laws have always been a good thing for the trapper as they have helped much towards making his occupation a lucrative one, but, as before explained, it is doubtful if it has ever, in any marked degree, tended to decrease the numbers of predatory animals.

It is true that continued trapping will cause the numbers of wolves and coyotes to diminish, but would not the trapping be prosecuted practically the same, even if there were no bounties? We believe that it would, for if the bounty offered were any great incentive, there would be more trapping done during the summer when the furs were of no account.

Neither do we believe that it ever induces others, not trappers, to kill these animals, for they will kill them on every opportunity, bounty or no bounty. It is man's nature to kill, for he is the enemy of all animal life.

# CHAPTER V

# HUNTING YOUNG WOLVES AND COYOTES

Of the many methods of hunting and otherwise capturing wolves and coyotes, employed by the professional "wolfers" of the west, none is more remunerative than the hunting of the young animals during the spring season. While the fur of the adult animals is of little value at that time and that of the young is not worth saving, the bounty which is usually paid for wolf and coyote pups will fully compensate for all loss from that source. At that time of year (March, April and May) there is very little fur of any value, to be had but the wolf hunter can combine wolf trapping and the hunting of the parent animals with the killing of the young, and the large bounties paid by many of the states and the various provinces of Canada, will alone enable one to do a profitable business.

In those parts of our country where the extermination of the wolves and coyotes is necessary for the protection of stock and game and the authorities and stockmen co-operate for the destruction of predatory animals, the hunting of the young animals during the breeding season should be especially encouraged. In no other way can the number of wolves be so surely reduced. To those who are well acquainted with the habits of the wolf, their time of breeding and the most favored breeding grounds, this mode of hunting is very simple.

Wolves breed much earlier than is commonly supposed, even by stockmen who have resided for a considerable length of time in the wolf country. The majority of young wolves are born in March in the Western States and the young of the coyote make their appearance mainly in April, but occasional litters of both will appear in May, and grey wolves may be born at any time during the summer.

On the western cattle range, the dens of the wolf and coyote are located mainly in the valleys among the foothills of the mountain ranges and among the low mountains, but seldom at any great elevation. The steep side of a hill or canyon facing the south is the most favored location, and the rougher and more broken and brushy the ground, the better it suits the wolves for denning purposes. They especially like knolls, strewn with large boulders, from which the male parent can watch for the approach of enemies.

As before mentioned, the mode of hunting is very simple. All that is necessary is to look carefully over the breeding grounds until tracks are found and these should be followed to the den. It is safe to say that at that time of year, nine out of every ten tracks will lead to a den. On the northern portions of the range, there is almost certain to be good tracking snow during the early part of the breeding season, but even if the ground is bare it is not generally a difficult matter to trail the animals to the den. A track that has been made in the evening should be followed backwards and one made in the morning should be followed forwards, as the wolves do most of their hunting at night and return to the den in the early morning. When the track can not be followed, if one can get the general course of it, the lay of the land will enable one, on many occasions, to locate the den.

Whenever the hunter hears of wolves, or their signs having been seen frequently, he should make a diligent search for the den. As the old mother wolf always goes to the nearest water to drink, the number of tracks at a watering place will often be a dead give-away and a careful search of the locality will usually result in the discovery of the den. As the den is approached, the tracks will become more numerous, and near by there will be well beaten trails. Where tracks are numerous one should keep watch for the male, sentinel wolf, as he will always be on the lookout somewhere near the den and his position will enable one to locate it more readily. As one approaches, the male animal will howl and endeavor to draw the hunter off in pursuit and thus prevent the finding of the den. Their tricks on such occasions show considerable intelligence.

When looking for dens on bare ground, a dog, if he understands the work is very useful. A fox hound that is well trained on fox is good, but if trained for this style of hunting especially, will be found to be better. Unless on the trail of a bachelor wolf, which by the way are occasionally found during the breeding season, the dog will readily trail the wolf to the den. It is best to go early in the morning as the trail will be fresher at that time and the dog is more apt to follow a fresh trail, therefore, more certain of locating the den. In all probability, one of the old wolves will attempt to draw the dog off for a mile or two, but in that case the mother will endeavor to return to her young. Sometimes they find it necessary to fight the dogs and try to keep them from approaching too near the den. Anyway the actions of the animals will show when they are in the vicinity of the den, which may then be readily located.

One hunter who uses a dog for this style of hunting says: "The kind of a dog needed is a good ranger, extra good cold trailer and an everlasting stayer. Then if he will only run a short distance after starting the wolf and come back and hunt the pups and then bark at them when found, you have a good dog that is worth a large price. There are plenty of dogs that will hunt and trail wolves all right, but very few that will hunt the pups."

The den is usually a natural one; a hole worn in the rocks by the elements, or in washed out cavities in the hard ground of the bad lands. Down in the valleys they sometimes den in the ground, enlarging the burrow of a badger or other animal. The opening is, as a rule, large enough to allow one to enter and secure the pups, but sometimes it will be necessary to dig the den open. For dens in the rocks, which are too small to allow one to enter, the hunter should provide a hook, something on the order of a gaff hook such as is used by fishermen. The hook should be of fair size, very sharp, and should be attached to a handle about three or four feet in length. A famous western wolf hunter in speaking of his outfit says:

"I will say to the boys who intend to hunt pups, get two or three strong fish hooks and a strong cord and carry them in your pocket. You can usually find a small stick or pole of some kind. When you find a den, tie your hooks on end of stick, wrapping cord very tight. If you use two hooks, put one on each side of stick.

Shove your stick in the den among the pups and turn or twist it and you will soon have a pup hooked. This works the best of anything I have ever tried; where pups are small. I have gotten many a bunch or pups this way, when my pick or shovel would be five or six miles away.

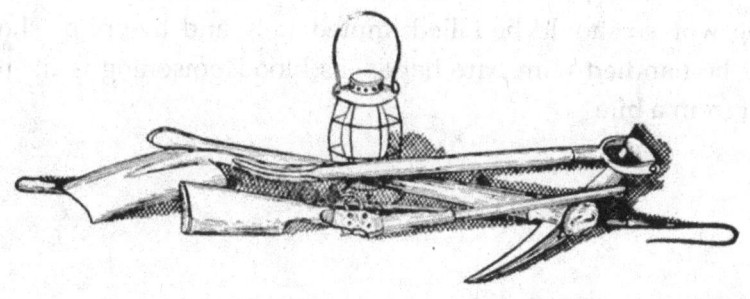

*The Hunter's Outfit*

When the pups get too large and strong to pull out alive, I put a candle on the stick, shove it into the den and lay on my stomach. With a 22 rifle I shoot the pups in the head and then they are easily pulled out with the fish hooks. I mean this for dens that cannot be dug out, as there are many of them in rock ledges and in holes in the solid rock. Instead of the candles mentioned by this hunter, some prefer to use a lantern and one "wolfer" uses a hunting lamp, attached to his hat. Some sort of firearm should be carried always. A revolver is good for use in the den, but a rifle is best outside.

It is not often that the mother wolf will be found in the den, as she usually makes her escape before one comes near, but should she be found at home she should be disposed of first. There is no danger, whatever, from the adult wolves. One of our western friends in speaking of this says: "I never hesitate in entering a wolf den, even when I know the mother wolf is with her young, and have never known one to act vicious, but always sneaking and cowardly. A few years ago at the Cypress Hills in Canada I entered a den and took ten pups. The mother crawled as far from me as she could and never raised her head. I set my 30-30 Savage and pulled it off with a rope, shooting her through the heart. It was forty feet from the entrance of the hole to where she lay, and it was midnight when I got her out. I had to move some dirt and rocks and it was a big job.

"I have killed other grown wolves in the den and have never known one to show fight. Of course, I always use a lantern to see what I am doing, and would not enter a den without one." The young wolves should be killed immediately and live pups should never be handled with bare hands, as blood poisoning is likely to result from a bite.

# CHAPTER VI

# HUNTING WOLVES WITH DOGS

Beyond all doubt wolf chasing as it is practiced in some parts of the country is one of the most fascinating of sports and in a place where the animals are fairly plentiful and the surface of the country is not too rough, is also profitable. In parts of the states of Minnesota, Wisconsin and Michigan, some of the professional wolfers use this method of securing their game and in the states lying west of the Mississippi River and east of the Rocky Mountains, also in Western Canada, wolf hunting is a very popular sport among the ranchmen.

Among the dogs that are most approved of by the wolf and coyote hunters, may be mentioned the fox hound, the greyhounds, and stag hounds of various varieties, the bloodhound and crosses of these dogs. The grey hounds are the swiftest of dogs and a pair of them are invariably to be found in a pack, the balance being some heavier and fiercer breed of dog, such as the blood hound, fox hound or a cross of the two. It is the grey hounds that run the game down and hold it until the arrival of the balance of the pack, the heavier dogs doing the actual fighting.

One who has followed wolf hunting extensively gives the following short but interesting description of the sport: "On the open plains of the west, wolves are often hunted with large swift running dogs, grey hounds, stag or wolf hounds or their crosses. The hunters go on horseback and the wolves are usually roused out of some coulee or draw. Sometimes trail hounds are used to start the game, on breaking from cover and being sighted by the running dogs the race is on. Wolf, dogs and horsemen, race across the often rough and dangerous ground at breakneck speed. The wolf, maneuvering to gain the coulee or cover of some sort and get out of sight of the dogs (the running dogs have only slight scenting powers and depend entirely on their sight). The lighter and swifter grey hounds, as a rule, are the first to overtake the wolf and by coming up alongside and snapping at his flanks, force him to turn and face them, thus giving the heavier and fiercer wolf hounds a chance to close in and grapple with and kill the wolf. Unless the dogs are well trained and very courageous, a large timber wolf often proves more than a match for the bunch of four or five dogs."

No matter what kind of dogs are used, they must be good tonguers and good fighters, and must have an abundance of strength and endurance. It is needless to say that the dog must be trained and this must be done at an early age. The young dog should never be run alone, for the wolf is likely to fight it off and once the young dog is driven back it will be spoiled for hunting purposes.

One of our Kansas friends in speaking of wolf dogs says: "We have plenty of wolves (coyotes) and have had for the twenty years we have kept dogs. As to breeding, we used an English greyhound bitch with courage, speed and a special hatred for a wolf, crossed with an English fox hound with all the qualities necessary, except the speed. We then picked the bitch with the most good qualities and crossed her with another fox hound whose ancestry is perfect. Here we get the dog we are using now and with which we have made the most satisfactory of catches. We seldom have a run lasting more than three hours and catch many, when vegetation is not too high, in from one to one and a half hours. Where this dog has the advantage over the fox hound is in speed and the fact that it is ever on the watch ahead for the game."

Evidently the party who used this breed of dog has endeavored to instill into the one type, all of the good qualities of the several breeds that go to make up the regulation pack of wolf dogs. It is surmised, also, that the one breed of dog is used alone, when chasing wolves. In Western Canada, wolf hunting is a favorite sport and one of the hunters from that section in speaking on this subject gives the following method of hunting:

"First, we put a box on the sleigh big enough to hold our dogs and then hook up a lively team, and strike across the country, leaving the dogs run along side. When a wolf is sighted, we get the dogs into the box and drive as close to the wolf as we can — that's usually from three to five hundred yards — then turn the dogs loose and cheer them to victory. The dogs usually run down the wolf within a mile, and we follow as fast as horse flesh can take us. When the leading dog gets alongside, the wolf stops, and in a second the dogs form a circle around him and he is a goner. Some hunters just turn the dogs loose, not knowing when they are ever going to see them again. That plan would not work with me. Good hounds are too expensive to monkey with that way. I have found that letting one or two dogs on a wolf trail spoils them, because one wolf will give two dogs all they can handle, and sometimes a little bit more, especially if they are young dogs. It takes two old dogs at least, to handle one wolf, and I have seen them get the hard end of it. The wolf perhaps would take to running into the scrub and then it wouldn't be long until a pair of wolves would be slashing your dogs or 'fleecing' the stuffing out of them."

Those who make a business of wolf hunting, or in other words, those who hunt for profit, do not always allow the dogs to fight and kill the wolf, but carry a gun with them, on all occasions and if they have an opportunity to shorten the chase by means of a well directed bullet, do not hesitate to do so. A high powered rifle should be used and one should learn to handle it in a business-like way. In the Western States where the large ranches are rapidly disappearing and the farm, with the barbed wire fence is taking its place, wolf hunting will soon be a thing of the past. Mr. Jack Kinsey, one of the most noted wolfers of the West, gives a description of an exciting wolf chase, in which he illustrates this point, and we give the story in his own words:

"While I was in Dakota last winter I had two exciting wolf chases. I was stopping with Mr. Wm. Clanton, a cowman, living seven miles south of Harding, S. D. One day I was in his shop putting a coyote hide on a stretcher, when one of his neighbors drove up and asked Mr. Clanton if he had a rifle. He said, 'Yes, there is a wolfer here who has one.' 'Why,' his friend said, 'there are two big grey wolves just back over that hill.'"

"I waited for no more but ran for my horse and gun. Clanton saw me going to the barn and told me to bring his horse. Now I was not long in getting those horses and we were soon on their trail. We followed their tracks about one and one-half miles when we sighted them. Picking out the largest of the two we both rode after him. The wolf started west towards some bad lands, but Mr. Clanton was riding a good young horse and he soon turned the wolf south, but now he was headed straight for a wire fence.

"Mr. Clanton would have succeeded in turning him again, but he struck a ditch full of snow, so the wolf got inside the pasture but I was fixed for wire fences. I had my trapping axe on my saddle and soon made a gate that we did not stop to fix up. We had run the wolf five or six miles by this time, and our horses were pretty well winded. So we pulled them up and let them take a slower gait until we got through the other side of the pasture.

"As I said before, Mr. Clanton was riding the best horse, so he kept the outside while I took advantage of the cuts. Mr. Clanton was just far enough ahead of me to make one throw at the wolf with his rope, but he missed him. The wolf cut in behind his horse, when I rode in front of him and put a 30-40 soft point in his head. He was a very large grey wolf. His hide stretched 6 1/2 feet long. On the way back we saw three more wolves and two coyotes."

We give the following spirited account of a wolf hunt which occurred in South Dakota:

"Will tell about one of my hunts behind a pair of wolf hounds that are certainly right when it comes to coyotes. I left my home here in Illinois on the 12th of December and arrived at Presho, S. D., early the 14th, where my friends met me, and we started for the ranch, which is about midway between Presho and Pierre.

"When we got to the reservation fence (Brule Reservation), we kept a lookout for coyote signs, and located a place that we thought would be all right, and planned a hunt for the following Saturday.

"The day proved all that could be desired, so we started out at noon. Earl, Claude, Mort, Chas., Sheldon and myself, with the two hounds, Ike and Lucy. A ride of about two miles brought us to the reservation, and the hunt was on.

"Our outfit consisted of our saddled ponies and team and buggy, and by standing up in the buggy seat we located two coyotes on the side hill playing in the high grass. A circle around the hill and Lucy discovered them and was off with Ike a second, as he was not as fast as Lucy.

"Away we all go across the prairie with the team and buggy following the reservation fence to keep the coyote away from the fence. It was a short chase, as Lucy soon had Mr. Coyote by the hind leg and turned him on his back quicker than it can be told, and Ike being close at hand soon had him by the throat, so by the time we could get out horses stopped and turned Mr. Coyote was no more.

"After skinning, we started for the buggy and Sheldon reported coyote No. 2 headed south down the draw, and Earl went after him around the hill and drove him back our way.

"A shout from that direction and the dogs have discovered No. 2 and we are away with Lucy in the lead, and this time we are not far behind, so that when the dogs got him we were right there, and the coyote not much hurt, so he gets a rope halter and is stowed away alive under the buggy seat.

"The dogs are panting hard and are very thirsty, with no water closer than five miles, so we head for home, but not far away on the hillside another one is seen and the buggy starts toward the left to head him toward the ranch, so the dogs will be running toward home when they jump him.

"This time Ike catches sight and is off, and Lucy cuts across to head him off. It is a short chase, for old Ike soon has his favorite hold and all is over.

"After skinning we started for home and as I hadn't ridden much for over a year you can gamble I was feeling pretty sore, for the pace a pack of hounds set isn't slow by a long shot. On driving into the yard the dogs were not slow about getting into the house and lying down.

"The live coyote we tied to the buggy wheel, and while I was gone after a strap and chain he bit the rope off and 'cut the mustard' for parts unknown with about a foot of rope still hanging to him.

"We have good hunting here in the spring and fall, plenty of chickens and, some ducks and geese, with lots of jack rabbits and

(Flicker Tails), prairie dogs, and their side partners, owls and rattlers.

"Our outfit is the bar circle outfit, O and I think our Holstein cattle are among the first herds in the state. Have since this hunt disposed of my interest in the O but still have a bunch of cattle at Presho, which supply the town with milk."

# CHAPTER VII

# STILL HUNTING WOLVES AND COYOTES

Hunting wolves with dogs, as described in the preceding chapter is certainly exciting sport but it is doubtful if it is as remunerative as still-hunting, especially in the rough sections where hunting with dogs is almost impracticable. In parts of the country where wolves and coyotes are plentiful, as they are in many of the thinly settled portions of the West, they may be still hunted at all times of the year. In the heavily timbered parts of the North, this method is practical only in winter.

The outfit that is needed for still-hunting in the West is one or more good saddle horses and the necessary equipment and a good, high powered rifle. A pair of field glasses will also be useful, but some hunters equip their rifles with telescope sights and the field glass is unnecessary. Hunters differ in their views, and with regard to rifles especially, there is a great difference of opinion. What one believes to be perfect, and which answers his purpose admirably, another has no use for whatever.

The arm selected should, however, have considerable power, and the flight of the bullet should be rapid, with a low trajectory. On the Western Plains the atmosphere is so light and transparent, and there is such a sameness to the surface of the country that one may easily be deceived in distances and with the high powered long-ranged rifle, there is less liability of errors, as the accurate estimating of distances is not necessary.

A gun of rapid action is also to be recommended and beyond all doubt the automatic acting arms are superior for shooting at running game. Personally, if the writer were selecting an arm for this kind of hunting, a high powered automatic rifle would be chosen, and it would be fitted with a small bead front sight and hunting peep rear sight. For use on horse back the shorter barrels are to be preferred.

In speaking of the outfit it is presumed that the wolf hunter would be a resident of the western country and would be hunting from home or anyway, making his headquarters at some ranch and hunting from there. If, however, he wants to go out into virgin territory, or if a stranger, he might find it necessary to camp out and in that case he would require a complete camping outfit. Some of the western wolfers use covered wagons for camps and this style of camp is very convenient as it may be moved easily, but if the surface of the country is very rough, this plan is not practical. In that case a tent would be needed and the hunter would use a pack horse in moving camp.

Speaking of saddle horses, in the more arid parts of the wolf country, the vegetation is scanty and horses require considerable time in which to rustle food. For that reason the same horse can not be used each day and one should have several so that each would have plenty of time to recuperate, after use. If one can obtain horses that will allow one to shoot from the saddle, so much the better. No special knowledge of hunting is required, but one should be expert in the use of the rifle, and should also be a good rider. All that is necessary is to ride over the rougher parts of the country, where wolves are most likely to be seen, and keep a sharp lookout for the game. It is always best to hunt to windward as one can approach closer to the game.

Where the bounty is sufficient to make summer hunting profitable, we would recommend this style of hunting at that time of year. In summer, hunting with dogs is not as simple a matter as in winter and trapping is not as good as during the colder part of the year. For coyotes, still hunting is a very successful method in parts of the country where the animals are plentiful and there is probably no place in which the method could be used to better advantage than in the sheep-raising district of Montana and Wyoming. There coyotes may be sighted every day and if the hunter would make a practice of following up the large herds of sheep to the summer range, he would always be sure of an abundance of game.

One is most likely to sight coyotes by riding along the coulees and over the rougher ground. About prairie dog towns are excellent places, as there they will frequently be found looking for the little inhabitants of the burrows. Other good places are the ragged, craggy parts of the Bad Lands and in the sage brush along the watercourses.

In winter one may follow the tracks in the snow and will stand a better chance of securing the game. While still hunting alone might not prove a very profitable method of hunting if one were hunting for bounty, it should always be used in connection with trapping and den hunting. As mentioned in a previous chapter one will often get shots at the adult animals near the dens and if one knows of the location of a den, he may often get a shot by watching it. Anyway the rifle should always be carried, and it should be used whenever a wolf or coyote is seen within range.

We will conclude this chapter by giving an account of a coyote still-hunt, as recorded by one of our western friends.

"It was one of those bright balmy September mornings, so characteristic of Wyoming, that I drove my horses down to water and noticed some coyote tracks in the mud at the edge of the water hole, and I decided then and there to have a coyote hunt that day. I was at the time in charge of a relay station, midway between two small towns and it was my business to look after the spare stage horses, for the stage driver changed teams here, leaving the tired horses in my care and taking on fresh ones. The northbound stage passed about 8:30 A. M., and the southbound outfit was due at about 6:30 P. M., which left me with practically the entire day at my disposal, to do with as I liked, and having my full quota of the spirit of our savage ancestors, I naturally turned to hunting the coyotes which abounded in that section.

"For some time past I had been doing practically no hunting. I say 'practically none' for I had not been out on a real hunt for several weeks, but I did have a short line of traps set and had been looking at them every second morning. On these trips over the trap line I always carried my 30-30 carbine on the saddle and had surprised and shot three coyotes, besides shooting at several more, one of which was wounded but escaped by crawling into a deep hole in a bad-land butte. Besides the three animals mentioned, I had caught in my traps up to that time, some twenty more.

"On this particular morning the 'Spirit of the Wild' called loudly, for as every hunter knows, there is something in the air of autumn which gets into one's blood at times, and there is no remedy except to go on a hunt. My trap line had been looked at the day before, so I was free for the day. Returning to the little sod house which I called my home, I got my rifle and six shooter, prepared a lunch and as soon as the stage had arrived, changed horses and departed, I mounted my horse and hit the trail for the hills to the westward.

"The section of the country to the west of the station was of the bad-land type, groups of buttes and ridges, radiating in every direction, seamed and honey-combed by the rains of centuries. While the country is very dry, the rains are veritable deluges when they do come, and the ordinarily dry water courses become raging torrents. Along these creek beds, sage and grease wood brush was abundant; in the hills, no vegetation was to be found. It was at all times a paradise for coyotes and occasionally a band of grey wolves strayed through those parts. However, the wolves had been rarely met with since the stockmen had abandoned the cattle industry and gone to sheep raising, but the coyotes had increased in numbers.

"At this time of the year, the sheep were being driven down from the mountains into their winter range and in addition to the coyotes which remained, throughout the summer, in the bad-lands, the still larger number which make a practice of following up the great bands of sheep were also appearing on the scene, and the day promised good sport.

"Riding westward about two and a half miles, I struck the bed of a stream and followed it up towards the hills. Here, I knew there were several prairie dog villages and about such places one is almost certain to find coyotes, so I turned my horse that way in the hope of getting a shot at one of the wary animals. My fond hopes were realized, for on rounding the hill at the edge of the first village I saw a large coyote slinking guiltily over the crest of the nearest ridge, but giving me no chance to draw the gun before he passed out of sight. Hastily riding to the top of the ridge, I saw the animal making his get-away down the draw at the other side and throwing my carbine to my shoulder, I caught a quick aim and fired just as he was rounding a spur of the ridge about a hundred and fifty yards away. Snap-shooting from horseback is uncertain at all times and on this occasion I had barely time to catch a half-hearted aim, so was not very hopeful regarding the results of my shot.

"Riding up to the spot, I dismounted and on looking the ground over, was elated to find a splotch of blood, but farther search revealed no other traces of the game. Naturally, I supposed that the animal had gone on down the draw and mounting my horse I rode slowly down the hollow, keeping a sharp lookout for the coyote. After looking the ground over for a quarter of a mile or more, and finding no signs of the game, I decided that this animal, anyway, was lost and returned to the scene of the shooting. Dismounting once more, I took the rifle and climbed to the top of the ridge to see what lay beyond. Imagine my surprise and delight when on reaching the top, which was low at this point, I saw the wounded coyote, vainly endeavoring to escape at the bottom of the depression on the other side.

"The first glance showed that the animal was badly wounded and could not last long, but fearing that it would fall into a hole, I took a hasty shot and had the satisfaction of seeing it crumple down, apparently lifeless. On approaching, however, I found that it still retained enough life to make a vicious snap at my hand, missing that member by only a few inches. As I watched it, undecided whether to shoot it again or leave it bide its own time, it breathed its last.

"It was a fine, large specimen and after skinning it, which required some twenty minutes of my time, I looked it over and found that my first bullet had struck it in the right hip, breaking the bone and passing through the body diagonally, emerging at the left shoulder. It was certainly a good shot and had I been using soft point bullet cartridges instead of full metal patched, the animal would have been killed instantly. It is surprising, however, how tenacious of life these animals are. The second shot had passed through the shoulders.

"I returned to the prairie dog villages but saw no more coyotes.

I did see a badger and fired at it just as it was about to enter the burrow, but missed the animal entirely. Going back to the creek bed I followed on up into the hills to a small alkali spring where I halted to eat my lunch. The water from this spring entirely disappears within two hundred yards of the place where it rises. The sun was shining fiercely hot by this time, and after eating my lunch I made a cigarette and crawled into the grateful shadow of the bank where I rested for a full hour. I had intended to make a large circle but found now that I would not have the time that such a trip would necessitate and so decided to go on northward through this range of hills and return home over the trap line.

"At the edge of the hills I found the traces of a sheep outfit and on rounding a spur so as to obtain a good view of the little valley beyond, I saw the white topped wagon of the herder at the far-side, but the sheep were farther down the hollow. Here I expected to find coyotes and I was not disappointed, for on riding through a patch of sage which covered several acres, a coyote broke cover on the opposite side. Three shots followed each other in rapid succession, throwing dirt and gravel over the fleeing animal but without harming him, and having no other effect than to increase his speed. I followed for some distance but failed to get another shot at the coyote and soon lost sight of it. Signs of coyotes were numerous here and about a mile farther I found the remains of two sheep which had been killed and eaten by the animals.

"As I rode over a small sag of a ridge and entered the head of a long narrow hollow, I saw a coyote trotting along down the draw about two hundred yards below me. The animal started to run before I could catch aim and I emptied the magazine in short order the last shot dropping the coyote, but it was not badly hurt and leaping to its feet it made off down the hollow. However, it enabled me to get quite close and putting spurs to the horse, I followed the animal, firing with my revolver. The third shot rolled it over and a fourth finished it, making two coyotes out of three shot at that day.

"Skinning the animal I mounted and hurried on to look at the traps. There were sixteen traps in the line and all but two of them had been undisturbed. Of these two, the bait was taken from one but the wary animal had apparently known just where the trap lay and had avoided it, the other held a young, female coyote. After looking at the traps, I returned home and dressed and stretched the skins of the captured animals.

"The skin of the coyote is of no value as fur, at that time of the year, but the combined state and stockmen's bounties aggregated $4 on each animal, so that I had $12 for my day's hunt. During the fall and early winter I captured by means of traps and gun, a hundred and thirty-three coyotes and four wolves. All of the unprime skins taken that fall were tanned by myself and made into robes."

# CHAPTER VIII

# POISONING WOLVES

Poisoning noxious animals is a common practice and is much used where the only object is to destroy the animals, and the finding of the carcass is of little moment, but the real hunters and trappers seldom resort to this method because of the large numbers of animals that are killed and lost. It is, indeed, a wasteful method of hunting as in all probability, three-fourths of the animals killed by the poisoned baits are not found until they have lain so long that they have become tainted, or ruined by mice and birds, so that both the bounty and the fur are lost. Anyway that is the conclusion of many of those who have practiced poisoning.

In many places where wolves and coyotes could be poisoned readily in early days the method is not a success at present as the animals have learned by experience to avoid the poisoned food. Strychnine is usually employed and this very bitter drug has a way of spreading through the bait, so that the wolf can sometimes detect it as soon as the bait touches the tongue. In such cases, the drug is never swallowed, but may be dropped on the spot or as is more often the case, it may be carried a considerable distance away before it is dropped. Again if the animal swallows the poisoned bait, it may be some time before it dissolves in the wolfs stomach and the poison begins to act, and if the wolf begins to feel the effects of the drug, it may start off on a run. In either case it is not likely to be found even if there is snow on the ground as the wind will soon obliterate the tracks.

In the government pamphlet before mentioned, Mr. Bailey has the following to say about poisoning:

"Many wolves are killed by poisoning, and more would be so killed if the methods followed were less crude. Strychnine is generally used with nothing to disguise its intense bitterness, the powder being either inserted in bits of meat or fat or merely spread on a fresh carcass. In most cases the wolf gets a taste of the bitter drug and rejects it, and if the dose is swallowed it may be too small to be fatal or so large as to act as an emetic. An old and experienced wolf will rarely touch bait poisoned in the ordinary way, but sometimes a whole family of young may be killed at a carcass. Usually when wolves are poisoned, they go so far before they die that if found at all it is not until their skins are spoiled. To encourage poisoning, it must be possible to secure the skins in good condition, or at least, to find the animals after they are killed, so that the ranchman may have the satisfaction of knowing that he has accomplished something toward the protection of his stock."

"In the use of poison it is of first importance to determine the amount that will kill with certainty in the shortest possible time. According to German and French authorities on toxicology, the smallest dose of strychnine that will kill a 25 pound dog is approximately one-fourth of a grain. Quadruple this for a 100 pound wolf and we have 2 grains. Mr. B. R. Ross, of the Hudson's Bay Company, found that this quantity would kill a wolf quickly. Experiments by Prof. David E. Lantz, of the Biological Survey, would indicate the best results from a still larger dose. One grain killed a 21 pound dog in seventy-five minutes, while 2 grains killed a 40 pound dog in twenty-seven minutes, without acting as an emetic. For a wolf, therefore, 4 grains of pure sulphate of strychnine would seem to be a proper dose."

"Tests on 40 pound dogs with 1 and 2 grains of cyanide of potassium in capsules caused the dogs to vomit in about fourteen minutes, after which they fully recovered. Other more deadly poisons can not be safely handled, and strychnine is the only practicable poison that can be recommended."

"For wolves, place 4 grains of pulverized sulphate of strychnine in a 3 grain gelatin capsule, cap securely, and wipe off every trace of the bitter drug. The capsules should be inserted in a piece of beef suet the size of a walnut, and the cavity securely closed to keep out moisture. The juice of fresh meat will dissolve the gelatin capsule, hence only fat should be used. The necessary number of these poisoned baits may be prepared and carried in a tin can or pail, but they should not be touched with naked hands. Old gloves or forceps should be used to handle them. The baits may be dropped from horseback along a scented drag line made by dragging an old bone or piece of hide, or may be placed on, around, or partly under any carcass on which the wolves are feeding, or along trails followed by the wolves. Partial concealment of the bait usually lessens the wolfs suspicion, while some kind of scent near by or along the trail insures its attention."

"The gelatin capsule will dissolve in about a minute in the juices of the mouth or stomach. When the strychnine is taken on an empty stomach it will sometimes kill in a very few minutes after the first symptoms of poison, and dies five or six minutes later."

Although this is the method recommended by the U. S. Department of Agriculture, it is our opinion that it would not be successful, for it takes too long for the gelatin capsule to dissolve in the animal's stomach, anyway that is the verdict of those who have tried the method. Those who have followed poisoning of foxes and wolves, prefer to place the poison in a small ball of tallow by making the tallow cup-shaped when cold, putting the strychnine inside and closing the opening by pressing the edges over it. None of the poison must be allowed to touch the outside of the bait.

The most common method of using the poison is to have out a large bait (the carcass of some animal that the wolves have killed is to be preferred), and after the animals are visiting it the poisoned baits are thrown about nearby. Any indications of the presence of man will make the animals suspicious and they will hesitate to approach the large bait but are very apt to pick up and swallow the small ones. We believe that this is the most successful method of using poison, but there are other methods recommended by woodsmen. Some place the poison in the large bait but we think this is entirely wrong as the wolf will taste the poison before it gets enough of the drug to cause its death. One party in speaking of the conditions in Northern Ontario, says:

"I think the wolf-poisoning business is being overdone. How would your readers like to find poisoned wolf bait within one hundred yards of their door, and some all round their houses within a radius of 300 yards? This bait consists in many cases of, say, half a deer. I think it is very wrong to allow strong, able-bodied men to pollute the country, now in the spring of the year, with large pieces, or in many cases, whole carcasses of deer. The wolf poisoner never stops to think what the result will be to his neighbor's dogs or poultry and cattle from their leaving large pieces of meat in an exposed position when the snow goes off. In fact, the thing is being carried in this part so far that neither fur-bearing animals nor fish will be procurable at any cost in a short time."

*Method of Preparing Poison Baits*

"As it is, the farmer's dogs have been suffering, and are nearly extinct here. For the last few seasons it has been quite common to see large quantities of dead fish round the lake after this poisoned meat goes into the water. I understood that parties laying out poison had to observe certain regulations, so that no person's property would be endangered thereby, and if I am rightly informed, it must be some person's duty to stop the nuisance. No doubt it is a good thing to get rid of the wolves, but the poisoning ought to be prosecuted far enough from settlements and from public roads that stock and poultry will not be endangered by the bait."

It is advisable when using poison to leave no human odors on the bait and to prevent so doing, some people prepare the baits without touching them with the bare hands. A simpler method is to make up the baits several days in advance and place them in a clean vessel, out of reach of all animals, and where they will be exposed to the open air. In this way, the human odor will pass away, and when they are placed out for the wolves they should not be touched with the hands.

The poison should not be placed in a capsule as that is too uncertain. There is very little danger of giving an overdose. The writer has seen strychnine used in large doses, considerably more than four grains, and the results were far better than when using smaller doses.

Strychnine is to be had in crystals or in crushed or powdered form. Both forms are equally deadly, but some poisoners claim that it will act more rapidly when crushed. The following article on poisoning is given by a Wyoming trapper:

"I have used strychnine and it is a poor idea for a trapper to use it as it destroys more fur than anything and also makes animals very shy about taking bait. The best way is to put the strychnine in lard which has no salt in it. I take some lard and put on my gloves or mitts and go in the shade or early in the morning where it is almost at the freezing point, so the lard will stay hard. I pinch it off in one inch chunks, take it between my hands and roll it in round balls, take a pocket knife or stick and drill a hole in it. Then fill it with strychnine and close the opening up by mashing the lard over the hole. Be careful not to get any strychnine on the outside as it has a bitter taste and if he tastes it he is almost sure to spit it out."

"Care must be taken not to touch the outside with the bare hands as the first thing a coyote will do is to smell it and if there is much human scent on it, he will not take it. I dip it in blood to kill some of the human scent. I have known a coyote to travel half a mile with a piece of strychnined meat in his mouth and then feeling it work on him, drop it and trot on."

"Now to show that it wastes furs and makes them wild, suppose it had snowed, blowed or covered his track in some way; a skunk in passing by and finding it, would get poisoned, which would be a loss of one fur, or suppose it were a mink or something else, it would have been the same. The chances are that the trapper would not have found it until it was spoiled."

"A coyote will travel a few hundred yards after taking the strychnine, even if it is put in lard or tallow. If he has eaten a large meal of frozen meat and then the poison, he will go far enough so he will be hard to find, and never found if you have any amount of bait to look after and it blows and covers his tracks up and maybe his carcass too. The result is that it makes other animals of his tribe leery about taking baits."

"Birds such as magpies, ravens, and crows will eat poison and fly off and die and be devoured by coyotes, foxes, mink, skunk, etc. The result is that many of them die, and what don't, get so sick that they are very careful about touching the next bait they see, generally giving it a wide berth. Then there are lots of birds, such as camp robbers and magpies that carry it off and store it away for martens or something else to eat and get poisoned by. There was a very large amount of skunk, but owing to the fact that many people poison whole carcasses for coyotes and wolves, they are rapidly disappearing."

"Some people claim that the dog destroys more furs than anything, but I believe that strychnine is a greater evil than a dog or at least in this county, so you see I have a good reason for advising a trapper not to use strychnine. It is the easiest way and the quickest way to get a few furs, if in the hands of an experienced man, but the furs are always a lower grade because the poison acts on the fur and there is always a hide now and then that the hunter will not find at once."

"The trapper will have the most furs and in the best condition at the end of the season and generally a place to trap more at the next season. The strychnine hunter will have to hunt a different place as what animals he failed to kill, will leave the country or become so wary that they will not touch bait. An animal which has recovered from a dose of poison, carries a pelt that is oft times worthless."

# CHAPTER IX

# TRAPPING WOLVES

After all of the various methods of hunting have been given a fair and impartial trial it will usually be found that trapping is the best means of capturing the wolf and coyote. Large numbers of traps may be set and attended to and the chances of making a good catch are greatly increased by so doing. If one has a liking for the work, makes a study of the animals and sets the traps carefully, good results are sure to follow. In all probability, four fifths of the coyotes and wolves captured in the United States are taken in this way. It is quite common for the professional trapper to take one hundred or more coyotes and wolves in a season.

The trap that is recommended for the timber wolf and the only one that was ever designed for capturing that animal is the "Newhouse" No. 4 1/2. It is a large and powerful trap, having a spread of jaws of 8 inches with the other parts properly proportioned. It is furnished with a two pronged drag and a heavy steel chain, guaranteed to stand a strain of 2000 pounds. The trap complete with chain and drag weighs about 8 pounds. A simpler and stronger chain fastening than that shown in the cut, is now used for attaching the chain to the trap.

No. 4 1/2 Newhouse Wolf Trap

Although the No. 4 1/2 is the trap recommended for timber

wolves, the No. 4 Newhouse is probably preferred by the average trapper, because of its lighter weight and its adaptability to catching coyotes, which are found in greater abundance than wolves. The trap has a spread of jaws of 6 1/2 inches, and its strength is sufficient for holding almost any wolf, providing the captured animal is not allowed to struggle too long, and that the trap is not staked, or otherwise securely fastened. Even when securely staked the No. 4 Newhouse will hold almost any one of the younger grey wolves, and it is mostly the young animals that are captured.

*The Two-Pronged Drag*

The standard trap is furnished with a short chain and ring, but when so ordered, the manufacturers will gladly furnish the traps with longer chains and the two pronged drag shown in the cut, or if desired the drags alone may be purchased and attached to any trap chain. The two pronged drag has an advantage over the four pronged kind, as it will occupy less space and may be more easily secreted.

It will be noted that the chain of the No. 4 1/2 trap has a double end. This is so that it may be looped around a small log or block of wood, if it is desired to do so. Some trappers prefer the chain without the iron drag, and for such the drag will be omitted. Others prefer to use the No. 4 trap with a 5 foot chain and a stone wired securely to the end. This makes a very good combination, but for some sections is not practicable as stones are "few and far between."

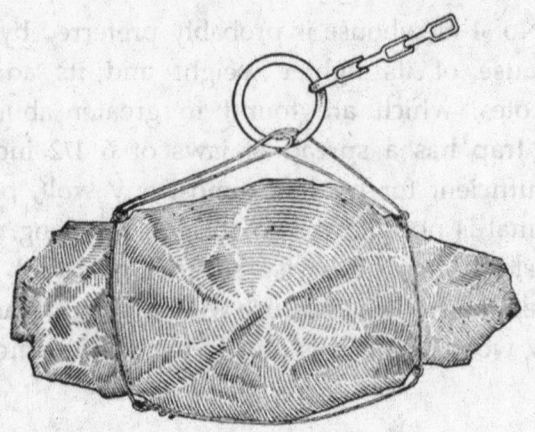

*Method of Attaching an Oblong Stone*

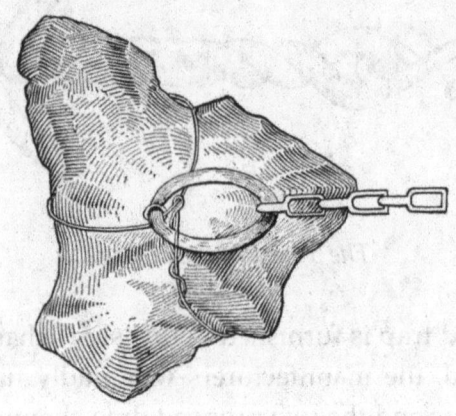

*Method of Attaching a Triangular Stone*

On the subject of fastening traps, Mr. Vernon Bailey of the Biological Survey gives the following:

"The best anchor for a wolf trap is a stone drag of 30 or 40 pounds weight, to which the trap is securely wired. A long oval stone is the best, but a triangular or square stone can be securely wired. Ordinary galvanized fence wire or telegraph wire should be fastened around the ends of the stone and connected by a double loop of the wire, then the trap chain fastened to the middle of this loop. A jerk on the trap tends to draw the bands together, and the spring of the connecting wire loop prevents a sudden jar that might break trap or chain. Twisted or barbed fence wire may be used if sufficiently strong, but it is not so easily handled. If no stones are available, or if the trap must be immovably fixed, it should be fastened with a twisted iron stake that can be driven below the surface of the ground. These stakes should be at least 18 inches long and of good iron straps three quarters of an inch wide and three-sixteenths of an inch thick. In light soil they should be still longer. See figures 1 and 2. If a picket pin sufficiently strong, provided with a swivel that will turn in all directions, can be purchased at the local hardware store, it may not be necessary to have a pin made to order."

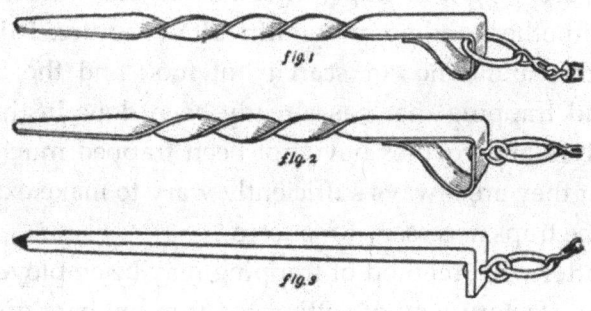

*Iron Stakes for Traps*

It is our opinion that the twisted pin would not be as satisfactory as the plain one shown in Fig. 3. If the swivel should lock, and fail to work, the stake might be twisted out of the ground by the struggles of the animal. With the heavy, square pin shown in Fig. 3, this could not occur. The pin should be made of wrought iron, about 5/8 or 3/4 inch in diameter at the top, and tapering to a point. The length should be the same as those described above.

The majority of the trappers who prefer to stake the traps use hardwood stakes and attach the chains by means of hay baling wire, twisting it with a pair of pliers. In many parts of the wolf country, hardwood is not to be had and many of the trappers use the spokes of old wagon wheels for trap stakes. We believe, however, that iron stakes are to be preferred to wooden ones.

The traps to use for coyotes are the Nos. 3 and 4 and the most suitable style of chain would depend entirely on the method of setting and fastening the traps. In some of the sets described, for both wolf and coyote, the traps, some three or four in number, are all fastened to one stake and for such a set the chains should be short, as also in the bank set. Where it is desired to use a drag of any kind, the chain should be 4 or 5 feet in length. This should be remembered when purchasing the traps and the method of setting that will be used should be kept in mind.

As before mentioned, most of the wolves caught are young animals less than a year old. After a wolf has reached its third year, it has attained a high degree of intelligence, and comparatively few of that age are caught in traps. In some sections wolves are more wary than in others and are more difficult of capture. This depends much on the abundance or scarcity of food and the amount of hunting and trapping that has already been done in that section. Where wolves and coyotes have not been trapped much, they are less shy but they are always sufficiently wary to make extreme care in setting the traps necessary for success.

No matter what method of trapping may be employed, there is only one satisfactory way of setting the trap, on bare ground. In a smooth, sandy spot, dig out a hollow the same shape as the set trap and of such a depth that when the trap is in place and covered with about 1/4 inch of dirt, the covering will be flush with the surrounding surface of the ground. A narrow trench may then be made, to accommodate the chain, and a hole in which to bury the drag. If a stake is used it may be driven under the trap and the trench will not be needed, or it may be driven at the side according to the method of setting but the stake must be neatly covered in all cases. The trap is then placed in position, the chain, drag and springs are covered and the portion outside of the jaws is filled with dirt, leaving only the jaws and pan uncovered. Now a sheet of clean paper should be placed over the jaws and pan and the whole covered with about 1/4 or 3/8 inch of fine dirt, covering the edges first and finishing with the center. A piece of canvas or hide should be provided, on which to place the dirt while setting the traps, and with which to carry away what is not needed for covering. When the setting is finished everything should look as smooth and undisturbed as it did before the trap was set.

*Trap Set and Ready for Covering*

In case the paper sags between the trap jaws and the pan, a few lumps of ground may be so placed as to support it, but care must be used so that no dirt gets under the pan. In freezing weather, make the nest for the trap somewhat deeper and line it well with sage leaves or some other light material, also fill in around jaws and springs with same, before covering. This will prevent the trap from freezing down. Do not put cotton under the pan as some advise doing, for if it gets wet it will freeze and interfere with the working of the trap. If the traps spring too easily, they may be remedied by drilling a small hole through the edge of the pan and inserting a tooth pick or small twig in such a way that it will support the pan. This will prevent birds and small animals from springing the traps. The same result may be obtained by bending the point of the "dog" or trigger of the trap upward and thereby causing the trap to spring harder.

Always before placing the trap in position, turn the springs towards the jaw that is held down by the trigger. This will allow the loose jaw to drop down to a level and let the trap rest more solidly in its nest. When adjusting the pan, always work from under the loose jaw, to avoid accidents.

Many trappers advise wearing gloves when setting or otherwise handling the traps, to prevent leaving human scent. It is our opinion that this is not only unnecessary but also useless, as the human odor will pass through a leather glove readily, and even through the sole of a heavy shoe. While there is no doubt that the scent of man will put any wary animal on its guard, there is no way to avoid leaving this same scent about the setting. This, however, will pass away after three or four days and it is after the traps have been set for some time that most of the wolves and coyotes are captured.

Although the traps may be handled with bare hands, we would advise that it is wise to not leave any more lasting odors than that from handling the traps, also do not leave any footprints or other signs of human presence. If, in summer, a line of traps can be set just before a rain, so much the better, as all odors and signs of disturbance will be removed by it. In winter a light fall of snow will have a tendency to improve the catch, as it will cover all human signs and to a great extent, smother the human and other odors that may have been left about the setting.

It is important that no lasting, foreign odors be allowed to remain on the traps or any of the implements used in making the set. The trapper should make it a point at all times to keep the traps clean and free from scents which might enable the animals to locate and avoid the trap. For the same reason strong smelling grease and oils, such as kerosene should never be used as preservatives, in fact, we think it best that the traps have no preparation whatever. Some trappers dip the traps in blood but unless the entire setting is saturated with same, it is not wise to do so as the wolf would be sure to locate it. When setting close to a large bait, it is well to rub the trap and chain with a piece of the bait, so that everything about the setting will have the same odor.

The same result may be obtained by covering the traps with hair from the animal used for bait, or with the contents of the paunch. When nesting the traps in sage leaves, as advised elsewhere, the odor of the trap will be greatly neutralized by the leaves, as they have a powerful odor. In trail sets on the cattle and sheep range, the traps may be covered with the droppings of the animals. All of these methods have the same result, namely, that of smothering the odor of the trap and allaying the animals' suspicions.

Just what will be needed for trapping wolves and coyotes will depend entirely on circumstances, but mainly on just how much of a business one wishes to make of it. The abundance or scarcity of the game, the nature of the country, the proximity to civilization and many other matters must also be considered. For the average professional trapper of the western cattle range, we believe the following will be about right: In the country lying just east of the Rocky Mountain Range, vegetation is rather scanty and as horses must pick their own living, they must have plenty of time for doing so, therefore, several saddle horses will be needed. In that way the trapper can change horses daily and give them a chance to rest and rustle food. For transporting the outfit and stringing out the traps, pack horses must be employed. Old, worn out horses will answer for packing and after the traps are once strung out, they may be killed for bait or otherwise disposed of, as one of the saddle horses may be used for what little packing is needed. The equipment should consist of a good easy saddle, bridle, pack-saddles, pack sacks, saddle blankets, hobbles, picket ropes, etc. If one is camping a good camping outfit will be necessary. Such an outfit would consist of a tent, blankets, cooking utensils, axe and some toilet articles. The average trapper would easily handle 100 traps, some trappers have many more, and the proper sizes and number of each size would depend on the proportionate number of wolves and coyotes found in that locality. The trapper must also have wire for fastening traps, stakes, paper for covering, a file for sharpening the axe and repairing traps, a whetstone, a pair of cutting pliers, a high powered rifle and plenty of ammunition, saddle scabbard, gun oil, hunting knife, axe sheath, etc. Such an outfit is costly, and is only useful to the professional trapper, but if game is plentiful, it will soon pay for itself.

For the ranchman, sheep herder or average western trapper, all that need be purchased is an outfit of traps of a number which may be conveniently handled, and a rifle with ammunition for same. All of the outfit that will be needed is to be found on any western ranch and as the trapper will not be camping out, the camp outfit would be omitted.

# CHAPTER X

# SCENTS AND BAITS

Scents for attracting animals to traps have been employed for many years, but trappers differ greatly in their views regarding its value. Some use scent only, to attract the animals, and make good catches; others use bait alone and condemn anything in the line of scent. Some use neither scent nor bait but depend entirely on "blind sets".

The value of scent for trapping wolves and coyotes depends on the kind that is used and the method employed, the time of year, the sex of the animal, whether trapping is prosecuted extensively, etc. We have no doubt that if the right scent is employed and used in the proper way that it will be productive of good results. In all probability those who are so ready to condemn scents have never used the right kind, or having tried the proper kind, have not used it in the right way.

If one will stop to consider just what scent is, and the object in using it, he must readily perceive its value, if the right kind is used. Scents are of various kinds and are expected to appeal to the animal in different ways. When one uses bait, it is the odor of same that attracts the animal from a distance, — why then will not a scent which suggests their favorite food also prove attractive? All animals of the dog family are very susceptible to food odors and the same scents will attract both the wolf and coyote. Then there are other scents which appeal to the animal's passions. These will be described in the following pages. They are especially attractive to the wolf during the mating season, but are also good at other times, and should be used without bait.

The habit of depositing urine on the same spot used by another for that purpose is characteristic of all animals of the dog family. This is sometimes taken advantage of by the trapper, and the wolf urine is used in that way.

In some parts of the country it is probable that one would be more successful by using bait alone; in other places blind sets would be better. For many localities it is best to use a good scent, and especially so at certain seasons. For trapping grey wolves in summer, it is especially valuable as at that time meat baits soon become tainted and are not attractive to the animals.

The United States Biological Survey have made exhaustive tests with scents and the result is given in the following:

"Success in trapping depends mainly on the use of scents that will attract the wolves to the neighborhood of a trap and keep them tramping and pawing until caught. Meat bait alone is of little use, for as a rule the wolves kill an ample supply for themselves. Many tests of scents, both prepared baits and various animal musks, have been made with wolves in the field and in the National Zoological Park. While some have given a fair degree of success, others have proved worthless, and no one odor has proved entirely satisfactory. Experiments are being continued, however, and new odors tried."

"Beaver musk (castoreum) and the commercial perfumery sold as musk have proved effective in many cases by causing the wolf to turn aside to follow the scented cross line and so walk into the trap. Siberian musk (from the Siberian musk deer) is very attractive to wolves in the Zoo. Oil of anise and oil of rhodium seem to have no attraction for wild wolves, and are scarcely noticed by those in confinement. Assafoetida is mildly attractive to wolves and coyotes at the Zoo, but used alone is very slightly, if at all, attractive to those on the range."

"Wolf urine taken from the bladder is used by some trappers, and is said to be very successful. It is bottled and kept until rancid and then sprinkled over the trap. The sexual organs of the female wolf immersed in the urine are said to add efficacy to this bait. The urine of the female in the rutting season is said to be especially attractive to males; it should be used in January or February."

"Fetid bait. — The bait that has proved most effective may be called, for lack of a better name, fetid bait, because of its offensive odor. It has been long in use in variously modified forms by the most successful wolf trappers, and its preparation is usually guarded as a profound secret. It cannot be credited to any one trapper, since no two prepare it in just the same way, but in most cases its fundamental odors are the same. It may be prepared as follows:

"Place half a pound of raw beef or venison in a wide-mouthed bottle and let it stand in a warm place (but never in the sun) from two to six weeks or longer or until it is thoroughly decayed and the odor has become as offensive as possible. If the weather is not very warm this may require several months. When decomposition has reached the proper stage, add a quart of sperm oil or some liquid animal oil. Lard oil may be used, but prairie dog oil is better. Add half an ounce of assafoetida, dissolved in alcohol and one ounce of tincture of Siberian musk, or, if this cannot be procured, one ounce of pulverized beaver castor or one ounce of the common musk sold for perfumery. Mix thoroughly and bottle securely until used."

The government has introduced this scent into Northern Michigan where it has been used successfully. Other very similar decoys are used extensively by Western trappers.

A scent which is highly recommended, and is used successfully by some Wyoming wolf and coyote trappers is made by chopping fine, equal portions of raw beef and fish and allowing same to decay in a covered vessel. After it is thoroughly decomposed, add an ounce of assafoetida dissolved in alcohol to each pound of the decoy. Animal matter of any kind should never be allowed to decay in a tightly closed vessel, as the gases may cause it to burst, but it must be covered so as to exclude the flies. The above scent is claimed to be very attractive to both wolves and coyotes and we know that the trappers who used it made large catches, one of them having captured over 200 coyotes in a single season. This trapper states that if the perfume of the skunk is added to the decoy, its attractive properties are greatly increased. This scent may be used in connection with bait, or without, as preferred.

One of the northern trappers recommends a scent made by chopping fine, equal parts of rabbit, skunk and muskrat flesh, with a couple of wild mice added, and allow to decay in a jar. The jar should be about 2/3 full and after it is decomposed a half ounce oil of anise and a quantity of skunk scent is added, and the jar filled with goose oil. This is the recipe as given but we can not guarantee it to be attractive.

Many of the old time trappers claim to have scents which will draw a wolf or coyote a half mile, or more, to the trap. Those who make such statements should always be regarded with distrust for the chances are that they only wish to sell the scent or the formula. In nine cases out of ten it will be found that the scents are worthless. One of our old time friends wrote as follows:

"I have tried several so-called patent decoys with very indifferent results. The only scent I care to use is the urine from a female wolf or coyote killed in running season; sprinkle a drop or two on bush, stone or ground near traps, but not on bait. After catching one coyote at a setting I never trouble to bait again as the urine and droppings will serve to attract other coyotes better than any bait. Have caught 6 at one setting, 5 of them with no other bait than the smell of the ground defiled by previous coyotes. They will come a long way to scratch and urinate on same spot, and seem to lose some of their caution."

Another successful trapper makes practically the same statements and his views are appended.

"After catching one wolf or coyote do not use more bait as the scent is strong enough to draw all that comes near. I do not use any patent decoy or scents, as I consider them useless for any game. The only scent I use is what I make myself and then only from February to April. In the summer I gather up four or five bitch dogs and as fast as they come in heat I kill them and take the organs of generation and pickle them in wide mouth bottles with alcohol enough to cover. I sprinkle a few drops on a stone or bush, stick in center between traps but use no other bait. This is also good for fox.

"The above method is the same as I learned it from an old Hudson's Bay trapper, Peirre Deverany, who was born in 1817, and had trapped all thru the British possessions and the Rocky Mountains and with whom I trapped for several years."

We find that many of the professional trappers condemn all scents except those which they, themselves, use, but as there are a number of successful ones using different scents it proves that there are numbers of good decoys.

"I use scent a good deal, but make it myself," writes one man who follows trapping continually. "The mating time is the best time to use it and the matrix from a female wolf in alcohol is very good to use. Put a few drops on a bone or stick of wood near the trap, say ten or twenty inches from it. If you have two traps set near each other, put the scent between them."

With regard to bait the grey wolf prefers horse flesh to beef. Colts are also preferred to old horses. It is the same as regards cattle; the calves and yearlings are invariably chosen. In the timbered sections where there is very little stock for the wolves to prey on, venison is perhaps the best bait. Antelope, jack rabbit, and in fact, almost any kind of flesh is good if the wolf is hungry, but the bait in all cases must be strictly fresh. Unless food is scarce, wolves seldom return to the carcass of a victim, but they do so occasionally and some are caught by setting traps in such places, especially in the North, during winter when the animals are hungry. If possible they prefer to kill their own game and it is that which makes the trapping so difficult.

The same baits that are recommended for wolves are also good for coyotes, but the coyote is not so particular regarding the condition of its food and will eat tainted flesh, greedily. They are very fond of mutton, prairie dogs, badgers and sage hens. As with the wolf, horse flesh is a favorite food. One of the southwestern trappers claims that they like fresh pork, in his section, better than any other food.

It will be an easy matter in almost any part of the country to keep the traps baited as the ranchmen and sheepmen are, as a rule, willing to furnish animals for bait. As a general rule, we advise the use of scent sets and blind sets in spring and summer and bait sets in fall and winter.

# CHAPTER XI

# SCENT METHODS

In sections where the wolves and coyotes can obtain an abundance of food, they do not care for meat bait and scent sets are recommended, especially for grey wolves. Such sets are also successful in summer when meat baits soon become tainted and lose all power of attraction. There are many ways of using scent, depending much on the kind that is used, and also on other things. One of the simplest as well as one of the best is the following:

Having found the route of travel of a band of wolves, one may be certain that he has found the proper place to set a number of traps, for the wolves are sure to come around that way again. A pass through the hills is an excellent place and as cattle, sheep and game animals are almost certain to be traveling that way at certain seasons, one is sure to find a trail of some sort traversing the pass. Having located such a trail find a spot where same is well defined and select a place for the trap, several feet to one side of the trail, where it may be placed between bunches of brush, cactus, rocks or any other obstruction that will guide the wolf over the trap. The obstruction must be a natural one as the wolf is certain to detect any artificial arrangement, and avoid it.

The trap should be fastened to a drag of some kind, which should be buried and the trap must be set and covered as explained in another chapter. In setting, the chain should be stretched out to its full length so that the drag may be buried as far as possible from the trap, and the disturbance of the soil is less likely to be noticed. The scent should be applied to the grass, weeds or ground at the back of the set, and so placed that in trying to reach it, the wolf or coyote must walk over the trap. It should not be placed too near the trap as the first impulse of the animal is to roll over the scent.

Mr. Vernon Bailey in his instructions for trapping describes this set as follows:

"The trap, chain and stone drag should be buried out of sight close to a runway, where the wolves follow a trail or road, cross a narrow pass, or visit a carcass, with the trap nearest the runway and flush with the surface of the ground; to keep the earth from clogging under the pan, the pan and jaws should be covered with an oval piece of paper and over this should be sprinkled fine earth until the surface is smooth and all traces of paper and trap are concealed. The surface of the ground and the surroundings should appear as nearly as possible undisturbed. The dust may be made to look natural again, by sprinkling water on it. Touching the ground or other objects with the hands, spitting near the trap or in any way leaving a trace of human odors near by, should be avoided. Old, well-scented gloves should be worn while setting traps, and a little of the scent used for the traps should be rubbed on the shoe soles. A piece of old cowhide may be used on which to stand and to place the loose earth in burying drag and trap.

"A narrow trail may be made by dragging the stone or scraping the foot from across the runway to the trap. A slender line of scent should be scattered along this drag mark or cross trail and more of the scent placed around the trap and 6 inches beyond it, so that the wolf will follow the line directly across the trap, stopping with its front feet upon it. With old, experienced and suspicious wolves, however, it is better not to make the drag mark, but to set the trap with great care, close to the side of the trail and put the scent just beyond it. If possible, place the trap between two tufts of grass or weeds, so that it can be readily approached from one side only."

Traps may also be set with bait and some scent used to advantage, in fact many of the decoys are to be used in that way. Two brothers who trap in partnership give the following method of trapping with scent and bait:

"This is one way of catching coyotes: We find all the horse meat we can, we even ask people if they know of any dead horses, or sheep or cattle. But the horse flesh is the best bait for them, then comes the sheep, that they like almost as well. Rabbits are also excellent bait for them, by putting a little pucky for scent along the side of the rabbit bait. We also tell how to make this 'pucky.'

"Secure all the small fish you can from three to four inches long. Trout is the best if you can secure them, but other varieties of fish will do; clean but leave the heads on, because you will find more oil in the head than in any other part. Cut them up so they will go into a bottle; stuff them in very tight, up to the neck of the bottle; then put a thin cheese cloth over the top of the bottle and let it stay there for about two weeks. It will begin to work good about that time, then cork it up tight, and in a few days it is ready for use.

"Now, taking the horse meat, sheep or rabbits, you have for bait, find a good place to lay the bait so the coyotes cannot get to it from all sides. Never make your trap stationary but wire the chain to a small log, a stick of wood about four feet long and three inches thick, leave a few knots sticking out on the log, and they will help tire the coyote out, by digging in the ground; wire the chain about in the middle, as it will drag harder for the coyote.

*Traps Set with Bait and Scent*

"Now dig a hole the same shape of your trap, where you want it to set, also bury the clog. Put your trap in its place and have it so it will be about one-fourth of an inch below the surface of the ground, not any lower. Put a piece of wool under the pan so birds and rabbits won't spring it; then take a piece of paper big enough to cover the jaws of the trap, take some dirt and put over the paper until level with the surrounding place, if anything a little sunken, just so you can notice it. Now set your trap about the length of the coyote from the bait and one more a little farther out, both in the same way as I said. Be sure and put your bait in such a place as to make them come around in front.

"If you have to set traps at a dead horse out in the open, put one just behind the hips, and one in between his feet where he lays. Set them as I have told you and you will get them. If you find a dead animal, that is, bait, I mean, also set your traps in triangle around him. Put your traps about one foot and a half from bait. Study them carefully and you will soon learn to set right. Try it.

"Some trappers say, don't let the traps touch your clothes, smoke and bury your gloves; and even say bury your shoes after each trip. We think all of this unnecessary for we tie our traps around us, wear warm German socks and overshoes, just as everybody else should do in winter. Set our traps with our gloves on or off, don't matter; when through, brush over with a small brush and leave it. Don't make any more tracks around your traps than possible. We made one freak of a catch, two coyotes at one setting in one night. One had a stub foot having been caught before."

A very good method is to find a large clump of cactus (prickly pear) with even, well defined edges, and set several traps near the edge and at varying distances. Use all possible care in setting, following the instructions given elsewhere. It is best to leave the setting some three or four days before placing the scent; that will give plenty of time for the human and other scents, that have been left there, to pass away and the ground which has been disturbed, will have taken on a smoother appearance by that time. Then go on horseback and saturating a lump of earth with the decoy, drop it in the center of the cactus bed. Do not dismount from the horse when placing the decoy. This is an exceptionally good set for coyotes. While they can not reach the scent, they will walk all around the cactus bed and are almost certain to step in one of the traps.

Another successful mode of setting is to place the trap in a trail where it leads through a clump of sage or greasewood and put some decoy by the side of the trail a rod or two away. The bank set which is described in another chapter may also be used without bait by placing some scent on the edge of the bank.

One of the Montana trappers uses this method: "Take your traps and boil them in lye water. Do not handle them with your bare hands but be sure and use clean buckskin gloves, and handle them as little as possible. Find a place where they run pretty regular, like an old road that is not used or a cow path or trail. Find a place that is sandy if you can, and set your traps lengthwise with the trail. Of course, you must dig out where you put your traps.

"Now cover your traps with a piece of deodorized paper and about one-half inch of sand. Get some water and sprinkle along the trail and over your traps to make it all look alike. You must not leave a lot of loose and lumpy dirt lying near your traps. Leave as little sign as possible. Wait two days before you go there again, and then go with a saddle horse and drop six or eight drops of good scent bait between your traps, and await results. Do not get off your horse when you go to put out the scent bait, for I know of no animal that is any more sly than the old grey wolf.

"Now I don't claim that this will work in all localities, but I have had fairly good luck with this set. I always use two traps at a setting for wolf or coyote."

# CHAPTER XII

# BAIT METHODS FOR WOLVES

Many of the sets used for coyotes are equally good for grey wolves, providing that one uses a trap sufficiently strong to hold them and almost any set that will catch the wolf is good also for the coyote, but there are some which are especially good for the grey wolf and we give here some of these methods.

One of the most successful is the following: Somewhere on the wolf's route of travel find an unused trail and selecting a well defined portion, set two traps close together as shown in the diagram. Have the jaws of the traps parallel with the trail so that there will be no possibility of the wolf's foot being thrown out by the rising jaws, and so arrange them that the pans will be about twelve or fourteen inches apart. The traps must be attached to drags of some sort, stones or iron drags, which must be buried, along with the traps. Great care should be used in setting so as to leave everything as nearly like it was before as possible. No loose dirt should be left lying about, and no tracks or signs of human presence should remain about the setting.

Two more traps should be set in a similar manner, somewhere on the trail, and from fifty to one hundred yards from the first two. The traps should be left setting some four or five days before placing the bait. This will allow all foreign odors to pass away from the setting. A large bait should then be placed midway between the two settings, and close beside the trail.

*Trail Bait Set*

On approaching or leaving the bait the wolves are almost certain to walk on the trail, and while they view all signs of disturbance near the bait with suspicion, they will be less cautious some distance away. In other words they will not be expecting danger so far away from the bait.

When looking at the traps, one should go on horseback and avoid dismounting near the traps or bait. In placing the bait one should, if possible, go on a wagon, or if more convenient, on a horse, and should drop the bait in place without stepping down on to the ground.

If desired a single trap may be used at each setting but as the length of step of the timber wolf is from eighteen to twenty-four inches, it is better to use two traps, for the wolf is likely to miss a single trap. The method will be found to work well in all localities and is as good for coyotes as for wolves.

Another very popular mode of trapping the grey wolf is with what is known as the square setting. This set requires four traps and they are arranged in the form of a square.

On a smooth sandy spot of ground, dig a hole about six inches deep and having attached the chains of all four traps to the stake, drive it in the hole until the top is below the surface of the ground. The traps should have the regular short chains and they should be arranged in the form of a square each about twenty inches from the stake. The traps must be bedded down, or in other words, they should be set in holes dug for the purpose as previously described and should be neatly covered. A narrow trench should be made for each chain and they must be covered also, so as to leave no sign. The bait should be fastened with wire to the top of the trap stake and the hollow beneath it may be filled with sand. The wire must not be visible and if a bird, rabbit or any small creature is used for bait, it must not be skinned or mutilated. When baiting with a piece of beef mutton, horse-flesh or the flesh of any large animal, it is best also to leave the skin on, as a skinned bait is likely to make the animals suspicious.

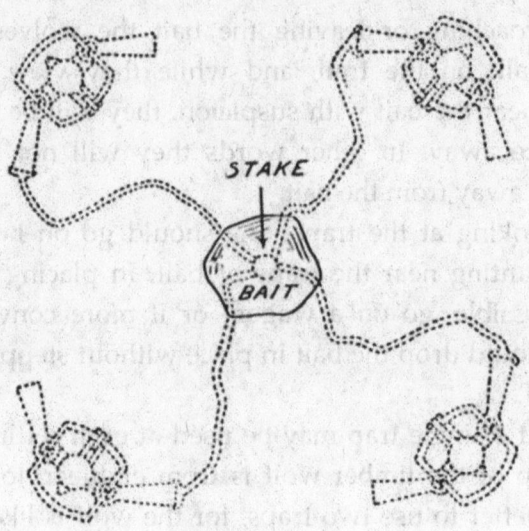

*The Square Setting*

If the animal's suspicions are not aroused, it will approach the set unsuspectingly and attempt to raise the bait, but when it finds it fast, it will step around some and is almost certain to step into a trap. It will be very likely also to land in another trap after it commences to struggle, and there will be very little danger of it escaping.

Many of the trappers who use this method use only three traps at a setting and arrange them in the form of a triangle. This is good but we believe that the use of four traps will give better results.

One trapper fastened his traps to iron pins, about 10 inches in length, and used this pin as a stake. The captured animal could easily pull up the stake but the entire bunch of traps would act as a drag, and it could not go far through the sage brush without getting fastened up.

One of the best methods for both the timber wolf and the coyote is what is known to trappers as the "cut bank set." All over the western country, along the water courses and wash-outs, will be found straight cut banks, sometimes overhanging. Select such a bank from 5 to 7 feet high, and if you can find two bunches of cactus, about 16 or 18 inches apart, on the top of the bank, this is the place to set the trap. If the cactus can not be found growing this way, place some there, being very careful to give it a natural appearance, so that it will look as if it had grown there.

The trap should be staked the length of the chain from the edge of the bank, and the stake driven out of sight. Set the trap about 20 inches from the bank, if for coyotes, and about 26 inches, if wolves are expected, and directly between the two bunches of cactus. Cover the trap nicely as per instructions on a preceding page, and fasten the bait between the cactus, on the very edge of the bank. When properly set, the animal can not reach the bait without stepping on the trap. When caught it immediately leaps over the bank, and as it can not get back, will be unable to make use of its strength in struggling, and will seldom escape. Another thing that speaks well for this method is the fact that the fur of the captured animal is always clean, which is more than can be said of those which are caught in traps set and staked on level ground, where they can struggle and roll in the dirt for hours, and sometimes days.

Mr. Ira W. Bull, official hunter for the U. S. Department of Agriculture, and now located on one of the Colorado forest reserves, writes as follows:

"It would be hard to make an estimate of the number of coyotes and wolves in this section, especially coyotes, as there are so many of them and they seem to be getting more numerous every year. There are not so many grey wolves, but still, enough to do a lot of mischief, as they kill stock and move on, hardly ever coming back to the carcass for the second meal.

"My method of trapping varies according to conditions and time of year. When I set with small baits, I first select a smooth open place, and cut a hole in the soil the size and shape of the trap. I set the trap in the hole and cover carefully, fastening to a stake or toggle, concealing by covering with dirt. I cut the bait in small pieces, probably 40 or 50 in number, and scatter around the trap, leaving everything looking as natural as possible. With a large bait, say the whole or half the carcass of a horse or other large animal, I set the trap in the same way, but use 2 or 3 or even 4 traps at the one bait."

An old time trapper writes as follows: "Water-sets are the best for wolves if the animals are cunning. The proper way to make them is to take a boat — don't walk along the bank but simply load your boat with lots of bait, such as beef head, shanks, entrails, or sheep that have died or have been killed by wolves. Start down the stream, looking for small sand or gravel bars lying just above the water and a few feet long. When one of these is found, run your boat up to it and leave a beef head, a quarter of mutton or such like, and then proceed on down to the next bar and bait it in the same way, keeping on in that way until the bait is gone.

"The wolf is very bait shy. It will take bait that it finds along streams more readily than on land. In a few nights after placing your bait, you will find that the wolves are working on it and have made trails down the bank of the stream to the edge of the water. You will observe that they all take the water at the same place.

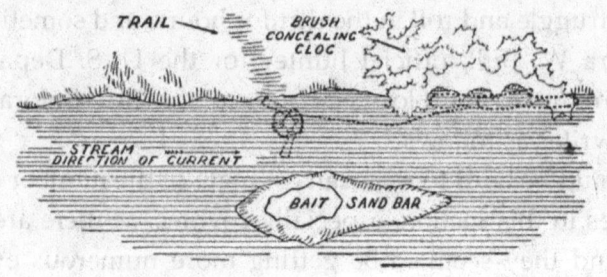

*Wolf Water Set*

"Now load your boat with plenty of bait as before, but this time take also a good supply of traps, the proper size for wolves, and a supply of clogs similar to fence posts. When you come to the bar, supply it again with bait. Fasten your trap to a clog, set the trap at the edge of the water in the trail and allow the clog to lie the full length of the chain, downstream in the brush. Splash water on the clog to wash it, and also on any brush you touch. Continue thus at the baited places, and you will be surprised at your catch, if you have never trapped that way.

"As for wolves getting scarce in the West, there are some places where the large wolves are decreasing. The coyote is becoming more plentiful every year. They are the worst of the two among sheep and small calves and colts. The sheep men on the desert are paying $40.00 per month to the trappers in eastern Oregon for wolves, besides boarding them and allowing them to keep the pelts. Some trappers are making as much as $150.00 per month. It is almost impossible to poison wolves in this country, but I can trap them successfully several ways."

One of the Minnesota trappers gives the following experience: "In the fall a man brought an old horse to give us for chicken feed, and after butchering it, we hauled the insides, head and feet out into the field along with some manure. After a few days we found that wolves were eating it, so when we butchered the next one, we dragged the insides around and put them in a little gulley and spread manure around; then set two traps, No. 3 Newhouse on both sides of the gulley and three traps down in the gulley near the bait.

"We set these traps on Monday, and on next Thursday father saw a fox running away from the traps, and found it had sprung one without getting caught. I think 4 or 5 wolves came around on Friday night, but they didn't get caught either. I moved one of the No. 3's nearer the bait, and on Friday father bought two No. 2 1/2 Newhouse otter traps. One of these we set where the No. 3 had been, and the other about six rods west of the gulley. We set the two No. 2 1/2 on Saturday morning. On Sunday morning on our way to church, we drove by the traps and found a wolf in the new No. 2 1/2 and a red fox in the No. 3 that I had moved up near the bait. These two traps were not over ten feet apart. On skinning the fox we found marks as if he had been bitten. It weighed 8 1/2 pounds and the wolf weighed 34 pounds.

"The wolves kept coming every other day. The next Friday we found another wolf in the same trap that the fox was caught in. On Friday of the next week we had another wolf in the No. 2 trap. On the next Thursday there was a wolf caught in the other No. 2 1/2 otter trap which was set six rods from the gulley, and that was the last one we caught up to February 15th. They don't seem to come around here now."

A Wyoming trapper submits the following: "I send you herewith a photo of a female grey wolf which I trapped in the spring of 1908; this wolf weighed 62 pounds. I caught her in a No. 4 trap, and when I got to within thirty yards of her I shot her with my 33 Special Winchester.

"The grey wolf is a powerful animal, and if a person goes too near them when they are in a trap they are apt to escape, and another thing, their feet are so large that a trap generally catches them by the toes. It is nothing uncommon for a single grey wolf to destroy $1,000 worth of stock in a year. This one that I trapped would have in a few weeks produced 12 cubs; just think of the damage which these thirteen wolves could have done.

"The grey wolf is hard to trap on account of being so powerful; they can kill a large steer or other ranch stock, in the shape of horses or cattle, and they like their meat fresh. I had fifty traps out and trapped 17 or 18 coyotes and several skunk while I was trying to catch this wolf.

"Here is the set I use: Find where the wolves have killed something or an old carcass, or find a trail that they are in the habit of using, for it is the habit of wolves to smell around anything they may find dead, and scratch around the same. Dig holes to fit the jaws and springs of your traps, put a wad of paper or wool under the pan of trap, and cover the entire jaws of traps with a piece of paper; then cover over the trap and chain with fine dry horse or cow manure, so that the covering will be level with the top of the ground, and make everything look as natural as possible."

"The accompanying photograph shows a wolf that I caught a few years ago and this is the way I caught it," writes one of the Wisconsin trappers. "First, I took the insides and stuff from a hog and placed it in a clover field and set three No. 4 Hawley & Norton traps around it, covering nicely with clover leaves, chaff, etc., but I guess I must have been a little careless, as a hungry wolf came along, ate what he wanted and scattered the rest of it around without springing the traps; so I thought I would teach him how to do that trick over again, and I took 4 more traps, making 7 in all, fastened one trap chain to the next trap, and in this way strung them out around the bait, fastening the whole to a logging chain that I had concealed under some clover seed hay.

"Then I covered everything very carefully with clover leaves, chaff, etc., and also some of the food out of the hog's stomach, as this food was smelling very sour by this time. I will also add that some of these traps were brand new, while some of them were very rusty, so I took first a new trap and then a rusty one, and set them alternately around the bait, thinking that this arrangement, together with the sour smell of the food, would confuse his nose a little, and I think it did, at any rate, in about a week he came back and got tangled up. He was caught only in one trap as his first jump would, of course, pull all the other traps out of position.

"He was a sorry looking specimen of a wolf, mixed up in all this hardware (seven big No. 4 Hawley & Norton traps and one logging chain), but we will have to excuse him as he "didn't know it was loaded." The best way is to fasten every trap separately, as in this way he may get caught in several traps, or more than one might happen to get caught at the same time, while if they hang together, he will not be likely to get caught in more than one trap, as in his first desperate struggle to escape he will pull the others out of position.

"I suppose it will make some of the old 'war horses' laugh to see that it takes seven traps at one bait to catch a wolf. This is the only time I have had as many as seven traps handy, so I thought I would fix him plenty. I generally use from one to three traps for each set, depending on surrounding conditions."

In addition to the trapping methods given in the preceding pages, there are many others used in the various sections of the country and all of them have some good points.

All trappers make it a point to set a number of traps about the carcass of any animal that has been killed by wolves, also animals that have met death through other sources. The trouble is that in open ground it requires so many traps to guard a large bait, and also the wolves become very wary and refuse to approach a large bait after one or two have been caught there.

For these reasons some trappers set their traps some distance away from the carcass, using small baits, and so placing the trap among clumps of brush and other natural objects that the wolf can only reach the bait by walking over the trap. Others set their traps without bait on any trails that may be found in the vicinity of the carcass, trusting that the animals will follow these paths when visiting the bait.

Some recommend dragging a large piece of bloody meat by tying to the horn of the saddle with a rope and setting traps without bait in the trail. Others set the traps in the same way and scatter small pieces of fresh bait all about.

Another style of setting which is sometimes used is to bury a good sized bait in a trail and set a trap on each side of the buried bait. All of these methods will give good results at times but one should never confine himself to any one method, as the animal will soon learn his tricks and refuse to have his toes pinched. It is wise also when using baits or scents to locate the set to the windward side of the animals probable course of travel as all animals can scent a bait at a much greater distance when passing to the leeward.

No matter what method is used, one must be a hustler and persevering. One can not possibly make a great success of wolf trapping unless he uses a large number of traps, and keeps them in working order and well baited.

# CHAPTER XIII

# SOUTHERN BAIT METHODS FOR COYOTES

If there are many methods of trapping the wolf, there are still more for catching the coyote, for it is a far more common animal, and while its range is over a smaller area of country, it is found in far greater numbers than the grey or timber wolf.

If there is any difference between the two, the coyote is more cunning and wary than the wolf, but the fact that wolves do not care for dead bait and the proportionately small number of the animals makes the capture of them more difficult and the catch very much lighter. All of the methods given for the grey wolf are good for the coyote, and in addition we give here the methods of various others, from all parts of the Southwest. The first is from Mr. Vasma Brown a noted coyote trapper of Texas.

"In the season of 1903-04, I commenced trapping about November 25th, and stopped about March 1st. I used seven No. 2 Victor traps, but consider No. 4 a better size. In the ninety-six nights that I trapped, I caught 182 coyotes, 4 skunk, 12 opossum, 3 coons and 12 cats. I only trapped for coyotes, but these other animals came along and got caught. Had I been trapping for skunk, opossum and wild cat, I would have caught about 200 of each, but their pelts were not worth more than 10 cents each."

"I took a piece of fresh meat and dragged it along a trail for about a mile. About every two hundred yards I set a trap. I scratched a hole in the ground just the size of the trap, put it in the hole and covered it up with a piece of paper and sprinkled dirt or sand upon it entirely concealing it. For bait, I cut some little pieces of meat and put about six or eight around the trap and then went on and set my other traps. I never failed to find two or three coyotes in my traps. My biggest catch in one night was six coyotes and one coon. I never use any scent. Fresh pork is the best scent that a person can use. I tie my traps to a log or a piece of brush."

"In the spring of the year, I have many calves and some hogs killed and eaten by coyotes. A calf about six months old, is the finest kind of bait for a coyote. A few years ago I saw a coyote kill a calf and as soon as I could, I put a 38 Winchester bullet through him. As many coyotes as I have caught in steel traps, I have never had one dig up my trap. They are very easily caught."

"In the winter of 1903 and 1904, I raced with a friend of mine catching coyotes. Our trapping places were about five miles apart, and there were just as many coyotes on his place as mine. I used seven No. 2 and he used nine No. 3 traps. He also used scent and I used none. When the day came to count and see who had the most hides, I had the most by one hundred and three coyotes, besides twenty-one other hides. He used scent and the animals were not very hungry. The scent they found on a suspicious place, made them shy, because they could not see what it was."

"If you will all think about it, it is plain enough. If an animal is hungry and sees a bait he is going to eat it whether it looks suspicious or not; and if not hungry and he sees a trap with bait, especially if he has ever been caught and gotten away, he will not bother it. The slyest of all animals have been caught in the most simple way a trap could be set, because they were hungry. My brother and I used to, and do now, catch coyotes at the carcass of a dead animal with our traps as unconcealed as you can set them. We always have had good success. We catch about one-third of what comes around."

"In the fall of 1895, my brother found the carcass of a dead horse. He set three No. 3 Newhouse traps at the carcass, and when it was all consumed, he had thirty-six coyote hides. His traps were set on the ground. A person could see them on the top of the ground at one hundred yards distance. The coyotes were very hungry and they cared for nothing but the flesh of the dead horse. They stepped everywhere, and on and in everything. Hunger makes an animal easy to capture."

The ease with which the coyote of the Southwest may be captured only goes to prove our statement in a previous chapter, that there is a great difference in the nature of the animals found in widely removed localities. It is certain that the coyotes of the

Northwest could not be captured in uncovered traps. There is also, as will be noted, a great diversity of opinion regarding the value of scent. In parts of the country where the animals take bait well, scent is seldom used and would not be as successful as in other parts.

"I trap on a small scale, but enjoy it more than any other occupation," writes a New Mexico trapper. "I intend trapping on a large scale next winter. It is mostly skunks, coyotes, badger and occasionally a fox and coon."

"Some of the brother trappers complained of wolves being hard to catch. There are very few here, but I can catch a coyote almost as easy as a skunk. I have caught lots of them. The best place to set the traps is on loose plowed ground or a sand bed, or anywhere you can conceal the trap so another person couldn't locate exactly where it is set. A cow trail is good. Setting it in grass is no good.

"I dig out a place in the ground just the size of the trap, and so the pan will be one-fourth of an inch below level, then set trap, put in place, take a piece of stiff paper (not too stiff), large enough to cover jaws, and place over jaws; then cover one-fourth inch with fine loose dirt. Brush the ground down smooth so it will all look alike. It is best to have two or three traps set at one setting, so you will catch him by more than one foot. When caught by one foot, they will soon sever their leg by jerking and twisting."

"Another pointer, when you catch a coyote at a setting, don't move your traps away, but set back in the same place, for the more you catch at a setting the better chances you have to catch more. I've caught six coyotes in one setting within the last two weeks. That isn't extra good but they are scarce here."

"By all means, don't pile up brush, stones, etc., around your traps when trapping for coyote. It doesn't make any difference how much the coyote scratches up the dirt, others will come back to the same place. I use most any kind of meat, such as rabbit, chicken and sometimes a coyote carcass until I catch one; then I seldom use any bait, for the scent left by the one caught attracts others."

Another trapper from Texas, writes, "In trapping for coyotes, there are three lines to be looked after with the eye of experience, viz: The where and the how of setting, and the bait. Beds should always be located between either hiding or feeding districts. In passing from one to another of these districts, coyotes follow in the main, the same route, and the experienced eye can soon locate a good place for a trap bed. These spots may be far apart or they may be near each other. The past season I had two not more than three yards apart. Failing to get such a location, the trapper traps by chance and catches by accident. After a rain the trap bed should be torn up and sunned awhile. Fresh sign is suspicious. Four traps make a good bed, but I seldom use more than two. Traps and chains must be well hidden and the ground left as level as would seem natural."

"If the traps are so fastened as to hold the animal to the spot, that bed is lost for that season. I prefer small pieces of worn out machinery, rusty iron, weighing ten or fifteen pounds for clogs. Then comes the most important matter of all — bait. In this I have deviated from anything I have ever noticed in guides. I use unrendered beef fat. Leaf fat is good, but I prefer what is commonly called gut fat. If taken off without too much care, it is best. Hung up and dried it lasts indefinitely. This dried article I cut up in pieces from the size of a pea to the size of the end of my thumb, the smaller the better, and scatter around over the trap bed, say 10 feet square. If the bed is in short grass, this baiting is better. The bait must be carried in a bag for the purpose, and must not be touched, in any case, until the traps are set. With traps well disinfected this is the slickest cheat I have ever seen worked on a coyote."

Some of the northern trappers will probably smile when they read of the following set, but the fact that it is used extensively in the South, proves that it is a good one for that part. It was contributed by one of the Arizona trappers. "There is plenty of small timber here, so the first thing I do when I find a good brush to wire the bait to, is to cut a drag about three to four feet long and about three or four inches in diameter about the center of the drag. Cut a notch on one side and in the center of the drag. Wire the ring of the trap chain securely to the drag in the notched place with about two lengths of hay baling wire. Lay the drag on the ground on one side of the pen and cover with brush. The pen would be brushed up all around about 18 inches high, except the entrance."

"I make a 'U' shaped enclosure about four feet long with bait wired to bush in the farther end. The pen should be about a foot wide inside of the brush. Dig a hole just inside the entrance of the pen for the trap, which set lengthwise, and cover even with the surface. Also be sure and cover the trap chain. Instead of cotton under the pan, I use a piece of canvas that just fits inside of jaws and put over the pan and cover all with sifted dirt from the hole until level with the surface. Place a stick across the entrance so that when the coyote goes up, if he wants to get the meat, he will have to go over the stick. It should be about eight inches in front of the trap. This keeps him from digging in the trap."

"Now fill in on both sides of the trap between it and the side of the pen with small brush or twigs so as to guide his foot into the trap. Do not put the twigs on the trap where you want him to step. I guide his feet right into my traps that way. Always lay the bait on the ground in the pen, wired to the brush or stake in rear end, as coyotes will not enter pen if bait is hung up. I use horse meat mostly, but sometimes rabbits and beef. Hawks and ravens are bad on rabbit baits, and cattle paw the traps up if set with beef."

"I set my traps from one-fourth to one-half mile apart, and use a fresh rabbit or fresh piece of meat and drag from one trap to the other, when making my rounds. Also spoiled fish scent is good for a trail. I never use gloves to prevent human scent in setting traps, and I consider it nonsense. After the first night a set has been out, almost any coyote will go into the trap. I use No. 14 Newhouse traps, and when they catch they never let go."

"I never set traps at a large carcass of a horse or other animal, for when one does when the coyotes come there to feed and one of their number steps into a trap, that generally settles it for the rest of them, and they will not come back. Set traps from one-eighth to one-fourth of a mile all around the carcass and bait with meat from the carcass is a good plan."

"I visit my coyote traps daily if possible, as they should not be left in the traps to frighten others away that would get caught, if the trap was set, and seeing that coyote in trouble, they will be very shy about coming up to the place afterward. I ride horse-back looking after my traps, and am able to get over a good deal of ground in that way."

Another coyote trapper from Texas gives the same method, and adds: "For bait take cracklings from either lard or tallow. Heat them in a skillet and when hot, cut up some garlic and drop in, but don't let it cook too much. Put the mixture in the pen the same as any other bait and see how it works. It does fine here, but it might be that there are so many coyotes here that they will eat any old thing. The best thing about that kind of bait is the buzzards will not bother it. I have tried it for coyotes, skunk and badger, and it is good for all of them."

This is the mode of trapping employed by a party from Southern California: "Now a word about trapping those cute little coyotes. As every one has his way of trapping for them so do I. The best way to catch anything that walks on four legs is to make a fool of them. Some people may think that is 'hot air', but I know better. The way to fool an old coyote is to take a fresh sheep skin and drag it, you riding, on a horse, for a mile or so in the hills near where your man is in the habit of going, (now be sure you do not touch it with your hands), until you find an open hill not too high. Have a stake there beforehand and have your traps set. The traps should be left lying in the sheep pen for a week before setting. When you get to the stake, hang your pelt on it, so when the wind blows the pelt will move."

"Mr. Coyote will be sure to find the trail you have made and will follow it until he sees the pelt, and then he will walk around it for a night or so, but he will not get too near the first night or for three or four nights, but he will be sure to get there after a while and try to pull the skin down, and he will forget about the traps and everything else, and he will be taken in just like all the other suckers."

# CHAPTER XIV

# NORTHERN BAIT METHODS FOR COYOTES

Mr. C. B. Peyton, who met such a tragic end, while attempting to arrest a party of game-law violating Indians last fall, wrote the following article on coyote trapping, several years ago. "I herewith submit my method for trapping the coyote, hoping it will be of interest to the readers. My outfit is as follows: Eighty steel traps, various sizes 2 to 4 1/2, two saddle horses, one short handle spade, one hunter's axe, a piece of canvas, some wool; 3 or 4 pounds of sheep or coyote wool is enough for one day's setting, one 30-40, 95 Model Winchester.

"When there is a bounty I do not start trapping until the frost is about all out in the spring. I start some morning with as many traps as I can set that day, four to the setting, five settings is a fair day's work if done right. I never bait until I have my entire line set."

"I have used the following style of setting with fair success, known as the square set. I select a spot where there is sand or no sod, cut a stake about 14 inches long, take four traps, fasten chains to stake, drive stake down until about two inches below surface, pull traps out about two feet from stake, a No. 2 trap chain is about right length, making a square set. Now dig out bed for each trap, placing dirt removed on canvas or blanket. Bed traps so there will be a half inch of dirt over them when covered; place enough wool under pan of trap to keep dirt out and keep rabbits or birds from springing them; leave a mark directly over stake to tell you where to place bait, when making your rounds with bait sack. Carry what dirt is left on canvas some distance, before dropping.

"I prefer a fowl for bait, such as an old dead hen, duck or grouse; place bait in center of setting on its side, lift a wing and drive a slender stake thru into the ground to anchor it and drop the wing down on top of stake to conceal it."

"Now back away a few feet and throw a few handfuls of dry sand or dirt over your tracks. If your work has been well done, it will be difficult to tell exactly where your traps are hidden, if your setting is properly located Mr. Coyote will not be slow to see or smell bait, as he is always on the lookout for handouts. He will take careful note of surroundings, if he sees or smells nothing suspicious, he will attempt to remove bait to some less exposed place and eat it or hide it for a future repast. He is very careful in approaching bait, making numerous circles of setting; if they succeed in reaching bait without stepping in one of the four traps, they soon find one when they attempt to raise the anchored bait, then begins a dance that lands him in two or more traps, there to await the coming of his friend, the trapper."

"Care should be used in killing captured animals, so they will not bleed on ground as that will spoil setting. I choke them with a small rope. Do not skin carcass nearer than 200 yards of setting. I use gloves always in handling bait or traps. I never go nearer than is absolutely necessary to see that they are not sprung. My line this spring, 1902, was 30 miles long. I went over it every other day, catching 43 coyotes in 6 weeks. I have never lost any coyotes by twisting feet off. When using square setting, they most always have two or more feet caught. I lose game and traps frequently by being lifted by human coyotes. I pull my traps up about the middle of May, then go to cruising after their dens."

The following method, submitted by another northwestern trapper is practically the same: "My outfit consists of the following: 60 No. 3 Newhouse single spring otter traps (I find they will hold any wolf and are easier to set than double spring traps), axe, 60 stakes 16 or 18 inches long, 12 or 15 pounds of wool or cotton, wool preferred, 20 stakes 10 or 12 inches long. A piece of oilcloth or canvas about 3 feet square, a light wagon and team, a 30-30 Savage rifle and four stag hounds. The hounds are trained to stay on the wagon until told to go, and will nearly always get a coyote when sent after him."

"In setting traps I choose a high knoll or a bare spot on the range and often the bed of a dry creek, where I see plenty of signs and then proceed as follows: Stick one of the small stakes where I want the bait and from 20 to 24 inches from it, I lay a trap and stretch the chain straight back, drive stake through chain ring and drive down below the surface of the ground an inch or more, then fix two more traps the same way at the opposite points of a triangle, set your traps and place a good wad of wool under the pan so that rabbits and other small game will not spring it, then proceed to bed the traps, and chains, placing all the dirt on the canvas. Now place your bait (I always use live bait if weather is not too cold but have had good success with dead bait). Lay an old dead hen or other fowl in the center and drive small stakes through it into the ground, firmly, cover the end of stake with wing or feathers of bait. Now step back and take dirt from the canvas and cover traps 1/2 to 5/8 inches deep, also cover your own tracks and brush over well with a brush. If traps are well set, it will be hard to tell where they lay. All dirt that is left on canvas should be taken away some distance and dropped."

"In using live bait, proceed the same way with traps, only bait should be tied by the feet with a good, stout cord and place a can of corn and one of water within reach of fowl, both cans to be set into the ground even with surface. Do not go nearer to traps than to see they are not sprung and do not shoot or club game in traps but choke to death with a copper wire on the end of a pole; a good stout cord will answer the same purpose. Wipe all blood off traps before setting again and brush out your tracks as before, and above all don't spit tobacco juice near your traps."

"Never set your trap by your bait; the bait is there to attract the animal," says a Colorado trapper. "When setting traps at your bait you only catch two or three, and by this time all the coyotes in the country have seen their comrades' doom at this particular place, and will stay clear of the place in the future."

"Find where there is a dead horse or cow in a draw, or some place where there are a number of trails leading toward it. Coyotes always travel on trails whenever they have the chance, in order to save their feet. Find where the trail goes thru some brush or high grass. Here is the place to conceal your traps, five in number, in the trail. Set them so they will take in eight feet of the trail, and there is no animal that can pass over these traps without stepping in one or more; fasten each trap with a pin eight inches long if the ground is frozen, and if not, the pin should be longer. If there is snow on the ground, put a piece of cotton under the pan and brush snow over them; if there is no snow, dig the trap down level with the ground, put a piece of paper over them and cover lightly with fine dirt. Use No. 3 Hawley & Norton or Newhouse traps."

"Use the same method for wolf; you need no bait for him. Find where he travels in a trail. He travels this trail every four or five days, take note of this and see if I am not right. Use No. 4 Newhouse or Hawley & Norton trap, with a heavy short chain and a good sized pin. When setting traps, take a piece of hide small enough to tie on the bottom of your shoes, and when within a hundred yards of the place where the traps are to be set, tie the hides on the soles of your shoes. Always use clean buckskin gloves when handling your traps."

"When you catch anything, move your traps a hundred yards or so, and reset. A coyote or wolf tears the ground up so that others get suspicious. If you have the chance to set traps horseback, take a hide and tie a rope on it, take this along, and when setting traps, throw this on the ground and step off on to the hide to set traps. When thru, get on your horse and pull the hide up with the rope. I learned this method from two of the best coyote and wolf trappers in Wyoming."

The following is from Joseph Casper, an Oregon trapper: "We have here, the coyote, wild cat, lynx, mountain lion and bear, but no grey wolves. Coyotes are plentiful, and I have seen as many as 6 or 8 at one time. A good way to trap them is by dragging the carcass of a sheep or pig through shallow ponds and set the traps in the water. The coyotes will follow the trail and will wade around in the water, looking for the bait. Traps may also be set by the side of some dead animal after the coyotes have been eating at it, or small pieces of meat or lard cracklings may be scattered around the trap. When setting traps on dry land, I would advise using some good wolf scent, to smother the human odor. I use the No. 3 Newhouse and No. 4 Hawley & Norton traps."

W. L. Williamson, a Montana trapper, in telling his experiences gives the following:

"Take some rabbits, chickens or other bait and make a drag out of it; dragging the bait from the horn of the saddle, and about every half mile, set two No. 4 Victor or No. 3 Newhouse traps in the trail and about 6 inches apart. Have a sheep skin to stand on when setting the traps, and do not step on the ground. Place all loose dirt on the sheepskin and after the traps are set and covered, get on your horse and lift the skin by cords, attached to the corners. Carry the loose dirt away from the setting."

This set is good for both grey wolves and coyotes:

"One day I went to the slaughter house, got a fresh cow head and took it about three miles away, placing it in the center of a small flat. I set several traps around it and the next morning I had a nice grey wolf, caught by two feet."

"When my father had his cattle down on our lower ranch, the coyotes killed a young calf one morning, so I took four Victor traps and set around it, and by 4 o'clock, I had two coyotes. I reset the traps and the next morning I had another one."

The trapping methods given in the following pages are from expert trappers of all parts of the central and northern portions of the coyote range.

"We have a $1.00 bounty on coyotes and $5.00 on wolves in this state (Wyoming) besides a stockmen's bounty in certain districts, ranging from $1.00 to $2.50 on coyotes, and $15.00 to $35.00 on wolves. I find the best way to find coyotes here is to go out in the open country where the sheep men run their sheep in winter, and when I can find a camp that has just been vacated by a band of sheep, I always figure on getting from one to five coyotes on that ground, as there is most always some dead carcasses left behind, and a good, dry place to set in."

"My method of setting is this, I have all my traps with the chains cut off to about six inches and a swivel on the end, and use a long iron pin about 5/8 inches in diameter. Usually, I take a part of a sheep with the hide on, and so place it as to leave but one natural way into it, where two traps put about ten inches apart will make it impossible for a coyote to get at it without being pinched. One can always find natural runways thru the sage brush, to make such sets."

"I also use the trail set a good deal, and always drag a piece of sheep pelt along from the pack horse. I use a pack horse most all the time, besides a saddle horse, and have two twenty-five mile circles out, with about thirty-five traps to each circle. In this way, I get from 75 to 150 coyotes every winter. The ground is too dry to freeze here, so I bury traps, pins, and use paper over and under jaws."

"A dead sure way to get a coyote every time is this, I can kill sage hens most any time, and always carry some on the pack horse. When it comes time to eat, first dig a hole to bury trap in and build a sage brush fire in it and singe a few of the feathers and some of the flesh in it, and set in the ashes. Who ever saw a camp fire that didn't have coyote tracks around it?"

"My way of trapping coyotes is to go to some prairie dog town and find an unused hole or one that has been filled up. Chop out a small hole two or three inches deep, then dig three trenches for the chains, then three holes for the traps, which must not be too deep nor too shallow. This requires practice and good judgment. They must be deep enough to allow the trap to be covered half an inch with dirt or sand, and still be even with the surrounding surface. Any deeper is too deep."

"Put a large piece of wool under the pan, and cover jaws, pan and all with a piece of heavy paper or light cloth, to keep the dirt from getting under the trap pan. Drive the stake with three traps attached until the top is two inches or more below the top of the ground; put the chains in their trenches and the trap in the holes dug for them. Cover all over with fine dirt the same as it was before being disturbed. Then take a brush made from stiff tough grass, a small brush or the wing of a chicken or sage-hen and brush out all finger marks, etc., then drop the last bait on top of stake and go away."

"The coyote or wolf will not come close enough to get caught the first three or four nights, but don't get uneasy, they will get bold after awhile, if you don't go too close to your trap when looking at it. When one gets caught in a trap set this way, he pulls to the end of the chain and swings around so as to step into another trap, then there is not much danger of him breaking a chain or pulling up a stake."

"In trapping the coyote or wolf, I make a bed some three or four feet each way, or nearly round. I set the traps after I swing the spring to the "dog" side. Then place the trap, say, about ten inches from the outside of the bed. Cover them with about three-fourths inch of soil. I cover the pan with a piece of gunny sack so the sack will be inside of the jaws. I place the pin in the middle of the bed, — everything is covered."

"I use bacon for my bait. After I have the bed all smooth and fine, I cut the bacon in very small pieces, then scatter them all over the bed, say some four inches apart. Coyotes like the bacon. They commence to pick up the small pieces and the first thing they know they are in trouble. I caught in two nights with the eight sets six coyotes."

"I make my beds near the cow trails. I have had better success making my beds near a dead carcass than to set the traps by the carcass. Last October we had an old coyote and five puppies that were killing sheep for one of our neighbors. I set one trap where the herder generally saw them. I caught the five young ones the first five nights. The sixth morning I went to the traps and they were dug up and the bait gone. I reset them and they were in the same shape the next morning. I said to myself, "Old girl, I will fool you." I made another bed some thirty feet from the old one. I set four traps in the new bed and fixed up the old one just the same as I had it before, only minus the traps. The next morning she was caught and had three feet in the traps. She ate all the bait on the old place and had pawed up the ground."

"I do not use scent. I have tried several kinds and consider them no good. I have trailed coyotes where they have been trailing my tracks and found them caught in the traps. I have set traps in the evening and found coyotes in them the next morning. I have been trapping coyotes and wolves for some five years in my county (Billings Co.) I am located on the Little Missouri River a short distance south of the old ranch that President Roosevelt used to own, what is called Bad Land Country."

"First boil your traps, and from the time you take them from the hot water, use gloves till set, gloves to be smeared with blood.

Take a pair of old shoes and nail on some blocks of wood cut from 2 x 4 stuff, the length of your shoes; nail them on from the inside of shoes with small nails, use gloves to do this. Now you are ready to start to where your coyotes are, so take four No. 3 or 4 traps and stake 3 feet long, with something to drive it with. Don't let traps touch your clothes while carrying them."

"When you get to your place that you have in mind, put stake thru all four rings of traps and drive down to the level of the ground; put your traps out each way so as to form a square, and bury each trap, chain and all. Make everything look as natural as possible. Put a small piece of wool or cotton under pan of trap and cover all well with dirt; take what dirt you have left from digging to set trap and carry away. Now leave your traps set till next evening, and then take a piece of beef liver or fresh hog lungs, put on your same shoes with blocks on and go put your bait in center of trap, (keeping gloves on), and don't expect to catch your coyote the first night, as he will likely come up close and take a look at things and go away again, but the third or fourth night, he will try to sample your bait, and when you catch your first one, the next one will walk in a lot quicker."

"I have caught as many as eight at one setting. Now mind you, in going to trap and resetting them, wear your shoes and gloves. I always bury my gloves and shoes in dirt to keep off human scent. I have caught lots of them this way, although, I have other methods. The main thing is to keep human scent off of trap and the ground where your traps are set."

"I saw a coyote jump over a sage brush about 6 rods from me one day, and shot at him as he struck the ground with No. 6 fine shot and killed him. As I went to pick him up, I found his hind foot in a No. 2 Newhouse trap. I took him out of the trap, took the trap, and followed his track for about one-half mile toward the top of the Butte, and found a dead horse. I left the trap, went back and skinned the coyote, took his hide over to Mr. Muma."

"About a week after killing the coyote, I went over the Butte, and found a man at the horse covering up some traps. I told him of killing the coyote and where to find the hide. He caught 11 coyotes at this horse up to February 1st. They set their traps from 10 to 30 feet away from the horse, between sage brush, where coyotes would be likely to walk in approaching the horse. They had eight traps set at this place, fastened each one to a limb about 3 feet long. I think they put some scent on the horse to keep the coyotes from eating him, as I did not see as they had eaten any of it during the time they had their traps set."

"I will give some good coyote sets, altho the season is about over now, March 8th, but some coyote trappers will trap most all summer in order to get bounty. I find that this thing in handling your traps with gloves on is all foolishness. Well, to begin with, take some lard cracklings, say a half dozen. Go to some brush where there is a trail going through, take your cracklings to the trail and scatter cracklings along trail, and set traps one at each end of brush in trail. This is a set hard to beat, boys. Another way is to find some old cow path, and if you see coyote tracks in this set a few traps along in it, cover traps, first spreading some brown paper over trap then some dirt. Take an old coyote foot, make tracks all around your trap, and you will have another good set."

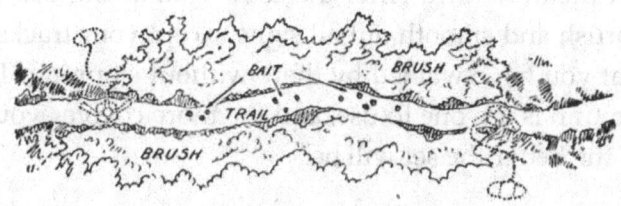

*A Trail Set*

"Here we have the coyote in larger numbers than any of the furry tribe, and he is here to stay, for his cunning is a match for the best of trappers, but many a one gets his toes pinched every season and his coat is worn the next."

"The best method that I know of to fool the cute chap is to find a carcass, and if they are feeding off it, then take about six or eight No. 3 or 4 Newhouse traps and set well back from bait. Set in trails leading to and from the carcass, but be very careful and leave no signs, for Mr. Coyote is very careful to look all around before partaking of his meal, and while making this tour of inspection (if you have your traps rightly and neatly set) he will get his foot caught."

"Never fasten the trap solid but to a drag so that he can drag it off and not prevent all the others from coming to the bait, and also he makes his hardest fight immediately after being caught, and if your trap is staked solid and happens to have a weak place, or your coyote is not securely caught, you are very apt to lose him."

"Find an old badger hole with a large pile of dirt in front of it. Take your traps, and everything needed to make the set with, walk straight up to the place and don't move out of your tracks while you set the traps. Put the bait, fresh meat of almost any kind, in the hole, so that the coyote can just see it. Set one trap about six inches from the mouth of the hole, a little to one side and another on top of the mound of dirt. Bury the toggles carefully the length of the chains from the traps and dig a hollow for the traps to set in. Be sure they rest solidly in their beds, so that they will not tip over if the coyote steps on the jaw. Cover neatly, with, first a piece of paper, and then fine dirt. After the set is completed, use a skunk's tail for a brush and smooth out all signs except your tracks. Have it appear that you have walked by there without stopping. The No. 4 Newhouse trap is the one to use, and the more coyotes you catch in one place, the better the set will be."

*Traps Set at Badger Den*

"Around most ranches are hollows, ditches, or strips of brush, along which the coyote approaches the ranch to catch chickens.

Along one of these places, about a quarter of a mile from the house is the place to catch a coyote."

"Take the entrails of a hog or other animal and go up the gulch until you find a place where the ground is loose and there is no grass. Set two traps about four feet apart and place the bait between, and about one foot from one of the traps. If the animal tries to eat the bait, it will be caught in this trap, and if it is suspicious and walks around the bait, the other one will catch it. Take a piece of the bait and erase all signs that you have made in setting the traps, so that it will appear that you have only come there to dispose of the bait."

"Look at the traps every other day, not oftener, and never go close to a set if it can be avoided. These may not be the best methods, but they are good ones, and I have caught many coyotes with them. When you get thirty or forty skins, you will think that they are well worth the trouble necessary to secure them, just to look at."

## CHAPTER XV

## BLIND SET METHODS

Where wolves and coyotes are plentiful and natural conditions are favorable, blind sets are very successful, especially for the wary animals that refuse to take bait. Conditions must be favorable in order to make blind set trapping feasible. There must be plenty of good clear trails traversing the country, and a comparatively rough locality will be found to be the best as, on rough ground, the wolves are more certain to walk on the trails.

It is only, perhaps, a small per cent. of the trappers who are able to make a success of blind sets, for it requires one who is very observing and a diligent worker. To make a fair catch requires that one runs a long line of traps, for he must depend on putting his trap just where the wolf will step, instead of decoying the animal into the trap by means of a bait, and no matter how careful he is in this matter, he is certain to set a lot of traps in bad places.

On the other hand, if food is plentiful and the wolves do not take bait well, or if they have become shy and wary because of persistent trapping, one is more likely to make a showing if he uses blind sets, in part at least. Then, too, he may be more certain of pulling in the "old veterans."

The reason that the blind set is more certain for the wary animals is that there is no bait to arouse the suspicions of the intended victim, and it is taken when completely off its guard. Such animals as the wolf, coyote and fox are always suspicious of a bait even though there is no trap there, and will sometimes steer clear of it for several days, simply because they think there may be something wrong there. They approach a baited trap warily and if they detect any disturbance or sign of human presence, they are off for good. With the blind set, that would not occur and if the trap is in the proper place, the trapper may be pretty certain of the animal when it comes that way.

In all parts of the wolf and coyote country, trails of some kind are to be found. On the Western Plains the stock trails are numerous and offer great possibilities for blind trapping. In the mountains, game trails are to be found and as such trails invariably lead through passes and other natural passage-ways they make excellent places for wolf sets, if on the animals' route. In the northern forests, moose, caribou, and deer trails are plentiful and good places for blind sets are to be found.

Main trails are the best always, unless one finds that the animals are traveling on the branches. The trail leading to the crossing place of a deep washout is an excellent place in which to set a trap. Unless the trail traverses a natural pass or leads to the crossing of a ravine, it is always best to be sure that the animals are traveling the trail before setting traps.

A narrow, well defined portion of the trail should be selected, and if there are bunches of brush, cactus or weeds on either side, so much better. A single trap may be used but as the animal is likely to step over it without springing, two traps are better. They should be attached to drags of some sort; either stones, chunks of wood or the pronged, iron drags. If the traps are staked the captured animal will tear up the trail and the next one that passes that way will stop to investigate and may locate the trap. With other sets, it is sometimes better to let the captured coyote or wolf scratch up the setting but with the trail set, it is best to use a drag.

A piece of canvas or cow or sheep hide should be spread on the ground and the trapper should stand on it while making the set, and should also use it as a receptacle for the loose dirt. A hole should be dug for each trap, the same shape as the trap when set, but a little larger, and of such a depth that when the trap is covered, the covering will be even with the surface of the ground. A narrow trench should be made for the chain and a hole in which to place the drag. The drag should be buried as far from the trap as the chain will allow.

The traps should be set with the jaws lying lengthwise of the trail. After filling in neatly with dirt around the springs and the outside of the jaws, a sheet of clean paper should be placed over the trap and covered with from one fourth to one half inch of fine dirt, covering the edges of the paper first to prevent it from sagging. When finished the whole should be brushed smooth and the surplus dirt carried away.

Sometimes one can find a long, deep ravine which is practically impassable to wolves and coyotes. At such places one may find small branches running out to the side and wherever there is such a branch, there is sure to be a trail at the first crossing place. Such a trail is sure to be used by the animals when traveling along the canyon for when they strike the lateral branch, they are certain to follow it to the first crossing place. That is the place to set a trap for them.

One of the trappers who is located on the coyote range of the Northwest, writes: "There are several ways of trapping for the coyote but none of them will hold good very long. The coyote will soon get on to the way you trap, and know as well where your trap is as you do."

The most successful way I have found is to take two No. 3 and No. 4 Newhouse traps and wire the rings together hard and fast. Set them in trails that are used by the coyotes. Dig a hole in the trails the right size for the traps. Double the chains up and put them under the traps, cover the traps lightly with dust, leaving everything as normal as possible. Two traps together make your chances double for a catch, and the loose trap answers for a drag. The coyote will not go far until he becomes entangled for keeps. I never use bait only to draw the coyote to some place where there are lots of trails leading in all directions. These trails I monopolize with traps as just described. I set it in the most likely looking place, then take a large sized bait, fasten it in a thicket in the vicinity of the traps, and your chances are good for a catch."

Another trapper gives his method in the following: "In setting traps for wolves and coyotes, I set them mostly on the trails made by stock. I use steel pins made from rake teeth. With a short handled axe I cut out a place in the trail so the trap will be level with the top of the ground when covered. I use paper over the jaws and set two No. 4 traps at a setting, putting them fourteen or twenty inches apart. A wolfs foot is good for brushing the dirt over the traps so as to make everything look as natural as possible. I use a pair of gloves in handling my traps and set them where the trail is narrow and on a little knoll, or where the trail goes around a bank or between two hills.

"Leave all wolf and coyote carcasses near the traps after skinning them, as they make a good decoy. A good plan is to throw your rope around a piece of meat and drag it from your saddle horn. Take a dozen No. 4 traps and go up and down the dusty trail and set them on the drag mark. If you hide them well, you will get Mr. Wolf or Coyote. I do not use bait in warm weather and not much in cold weather. A grey wolf is hard to catch by bait, unless very hungry and he is seldom troubled that way where there are cattle and horses on the range."

# CHAPTER XVI

# SNOW SET METHODS

When the ground is covered with snow, trapping for wolves is exceedingly difficult and there are few, if any trappers who can make a success of it. Throughout Northern Minnesota, Wisconsin and Michigan, as well as in Canada, a few are caught by the most persistent hunters, but the winter catch never amounts to much.

It is difficult to make a set in the snow and leave no signs when the set is finished, and even if one can make a neat set it will seldom remain long in working order. This is the rule, but there is one exception, a set which is successful, but can only be used in places where the winter temperature is such that the snow will remain a long time in a loose, powdery condition. In other words it can only be used successfully in the North, where the weather is very cold. The method referred to is the one used by the northern Indians for trapping both the fox and wolf. It is made as follows:

Having the trap attached to a heavy clog, and well cleaned by boiling or washing, go out onto the ice of some windswept lake and scrape up a pile of snow. Make it cone-shaped about three feet in height and six or seven feet in diameter at the base. Bury the clog, or drag, in the mound, and stretch up the chain, so as to bring the trap to the top. Make the mound hard by beating it with a snowshoe, and in the top, scoop a hole about five inches deep and somewhat larger than the trap. Line this hole well with dry moss or cat-tail down, the down is best, and place the trap in the nest. Fill inside of the jaws, and under the pan with cat-tail down and after the trap has become cold, so that there is no danger of the snow sticking to it, sift snow over it, to the depth of an inch. Do not touch this snow with the hands or it will freeze hard and the trap will not spring.

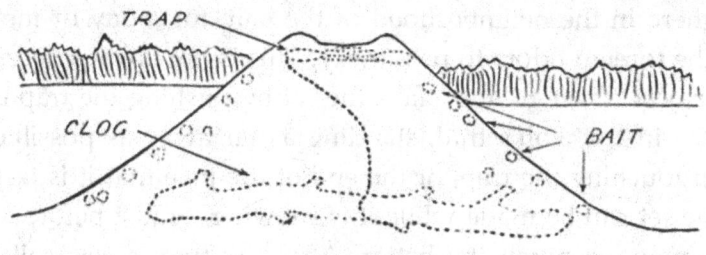

*A Snow Set*

The bait should be cut into small pieces and tucked into the sides of the snow mound, where it will be out of sight of the birds. Brush out your tracks as you go away and the wind will soon erase every vestige of signs, leaving the snow as smooth as it was before the trap was set, but the mound will freeze hard and no amount of wind can drift it away.

Such a set will remain in working order as long as the weather stays cold. A fresh fall of snow will bury the set for a short time but the wind storm that always follows a snowfall will blow all loose snow off the mound, leaving just a sufficient amount over the trap, as that will be sunken somewhat below the level. The human scent will also pass away in a short time.

This set is practically the same as setting a trap on the level and scattering the bait about, the only reason for setting it on a mound being that it will not be buried by the falling snow.

While the set described does well where wolves are making an occasional trip across the country, for places where the animals are plentiful, some other methods must also be employed.

If one can find where the wolves have killed some animals and are feeding on the carcass, he will note that they have trails where they approach. One may put out a large bait and they will beat a trail about it at times. These trails make excellent places for snow sets.

The best way to make the set is to fasten the chain of the trap to the end of a long pole clog, and having set the trap, split the end of the pole and pinch one of the springs in the split. Now slip a clean paper bag over the trap and stand the pole and trap against a tree somewhere in the neighborhood of the bait, for a day or more, to allow the foreign odors to pass away. This is not always necessary, but it is best. Then go and make the set by pushing the trap under the snow in the wolf's trail, standing as far away as possible and without touching the trap, or the end of the pole that it is fastened to. If the set can be made while it is snowing, or just before a light fall of snow, so much the better. After the trap is set walk back stepping in the same tracks and brushing the footprints away with a bunch of evergreen boughs tied to a stick.

This set is good if the wolves are visiting the carcass regularly but will not remain in good condition very long, as a heavy snowfall will put the trap out of commission.

One of the professional wolf catchers of the western mountain regions, gives the following set: "When there is snow, I cut a piece of soft cloth, white preferred, the size of the jaws, when open, and lay it over the trap, being careful not to let it get into the corners, next to the springs; then cover lightly with snow. The cloth will prevent the snow from getting under the pan of the trap and thus prevent it from springing. It is also a good plan to put a brace under the pan, so that the birds cannot spring the trap. A small forked willow will do, but a better plan is to drill a hole through the pan, near the edge, and place a match, or a tooth pick slanting through the hole to the bed of the trap."

"I use the No. 4 Newhouse trap with long chain, for coyotes and wolves. The bait, I cut in small pieces and scatter all around the trap."

One of the coyote trappers from Saskatchewan, Canada, says: "I will give a snow set for coyotes that an Indian showed us and we proved its merit. Select a good hard snow drift, set your trap and lay it on top of the drift, then with a knife, mark the snow around the trap, remove trap and dig out the snow to a depth of three or four inches, replace trap in hole so that the pan will be about two inches below the surface. Now go a little distance off and cut a cake of snow large enough to cover hole, in which lies the trap and scrape it as thin as possible without breaking. This requires care. Now place the cake over the trap and sprinkle some snow around the edges so as to leave all smooth. The chain and clog of course, should be well buried in the snow."

"I have caught a coyote in a set like this after a big storm, the snow having blown clear over the drift and not injuring the set in any way; all I did for bait was to set my trap by a little bunch of grass. Of course, it is evident a set like the above will only apply when it is cold and there is no chance of a thaw. Another important point to be remembered in setting traps is to give them a firm bed. When a trap is sprung it kicks back the same as a gun but when on a firm bed it has the greatest chance of a high grip."

In portions of the North, snow sets are used considerably. The sets given here were sent by a Minnesota trapper who claims to have used them successfully.

"I have trapped wolves a good many winters in this part of the country, but they are very scarce here now. As to my way: I use a No. 4 trap and set under the snow. If I can find a place where their paths come together or cross, I select it as a favorable place for catching them. If there are a couple of bushes near together with the paths between, I set my trap there, pushing it under the snow from a couple of feet back of the path, taking care to make as few tracks myself as possible and to fill those up and brush with a bunch of twigs or weeds for a distance of twenty feet or more. I sift snow over the trap also and leave everything as natural as possible. This method I have found very successful in capturing these shadowy pests of the prairie."

"When ponds, lakes and rivers are frozen over and the snow is deep, wolves are apt to travel on the ice; any dark object out on the smooth expanse of snow on lake or river will at once attract their attention and they are apt to go and examine. A crow, rabbit or bait of any sort; let it be up where it can be seen at a distance. Place two or three traps around the bait at a distance of three feet, put pieces of white paper, one under and one over the trap, then cover carefully with dry snow by sifting it with a piece of wire screen."

"When traveling an old trail or timber road thru the woods, reach out to one side as far as possible and place a piece of bait with some of the scent on it or near it, and place two traps half way between bait and trail, also one directly in the trail. Set and cover it as on the ice. It is a good plan to scatter a few beef or lard 'cracklings' along your trail. No. 3 traps are about right for wolves, and the No. 2 1/2 Newhouse otter trap makes a good wolf trap if the attachment is taken from the pan."

"To sum up, the trapper who makes a success of trapping wolves must make a study of it and must often contrive methods suitable to his particular trapping grounds."

The following extract from a letter received from a Canadian trapper, tells of a very successful coyote set.

"One day I found a dead sheep in the pasture, and dragging it down to the edge of the lake, I set my traps around it, covering them nicely with wool from the sheep. I told the boys I would have a coyote in the morning, and so I did. On the second morning I had a red fox, on the third morning a coyote, on the fourth a fox and on the sixth morning another coyote. Then I did not get any more for a week from which time, I caught one now and then until spring. I think I caught 23 coyotes and 2 foxes at that one bait. When the snow got deep, I set the trap on top of the bait. When a coyote came along he would smell the bait and would dig down through the snow, into the trap. I wore skis when looking at the traps and never turned around near a setting."

"My last winter's catch was as follows: 69 coyotes, 5 lynx, 2 red foxes, 5 badgers, 12 weasels, 12 muskrats and 2 mink."

"I want to tell you how I catch coyotes," writes a North Dakota trapper. "I set two or three No. 3 Victor traps around some dead horse or cow, cover the trap with a piece of paper or cheese cloth, then throw snow over that, having it look as near like the surroundings as possible. Sometimes I use a fresh beef head, but the coyotes are so shy they will not go close enough to get in your trap for sometimes a week, unless they are starved to it."

"I think the coyote is as shy as most any other animal. I do not think they can smell the steel traps for the strong smell of the fresh meat or carrion but they are afraid of your tracks, and naturally suspicious of everything. When I first tried to trap coyotes, I drove up within a few rods of where I wanted to set my traps, went and set them, and did not pay any attention to destroying my tracks. I would never catch any until snow filled up my tracks."

"Now I set my traps off of skis or snow shoes or drive up close to where I want to set my trap, and drag some fresh meat over my tracks; they are not afraid of a sled track for they will travel for miles in sled tracks when the snow is deep."

We will conclude this chapter with an article written by a Canadian trapper, telling how he caught his first coyote:

"This is my second winter in Alberta and I must say that we are having one of the good old fashioned kind. The snow is over two feet deep on the level, and the thermometer on one occasion, went on a strike. It was only 36 degrees below zero this morning.

"Last winter, which was very mild, was a poor year for catching the sly old coyote. He was too well fed and could get around so easily that he never suffered the pangs of hunger, so was constantly on the watch for danger. We had a cow that committed suicide by falling into the manger, and I thought she would make good bait. So she did until I set some traps around her and from that time the coyotes would come and look at her, but would not venture near. However, I succeeded in catching three large dogs.

"On January 5th, I changed my boarding place, moved to within a half mile of Battle River and Lake. The coyotes were quite numerous around the lake and river, and made nightly excursions up around the buildings, feeding on a dead horse, cow or calf. The boys had a couple of traps set beside a cow, but the cattle would spring the trap while feeding at the straw stack where the dead animal was. Then I took a hand and set the traps on runways used by coyotes. I set them with great care, but all I found was a footprint about two inches from the pan of a trap. Sometimes they would go as far as the trap and would turn around and retrace their steps. One night they actually scratched the snow off of the trap, as if to show me that I needn't try to fool them because they were on to my game.

"However, my turn came. There was a little old straw pile that they seemed to like to run onto, to see if the coast was clear. There I set a trap, covered it and the drag nicely with snow, brushed out the tracks with a twig and made some nice tracks right over the trap with an old coyote's foot. I also threw a little piece of meat up on the stack.

"Friday morning I ran down to my trap and was surprised to see it gone. I saw some blood on the snow but could not realize — no doubt on account of so many disappointments — that there was anything in the trap. However, I followed up the trail and you can imagine my delight in finding a big, fine, dog coyote in the brush. The next thing was to kill him, and I assure you that they are the hardest animal to kill with a stick an inch in diameter that I ever tackled. I pounded him on the head until his skull was crushed and still he breathed.

"On Sunday morning I took a walk down to a trap I had on another straw pile and when within a hundred yards of the stack I saw a coyote rise up, take a look at me and then start to run. I ran, too, and when I arrived at the other end of the stack there he was fast in my trap. I thought that was pretty good for I had actually chased him into my trap. Two coyotes in three nights was pretty good, with only three traps, and I was quite proud of myself, but that was a week ago and number three only came last night. I am in hopes of more before spring, but never will I have the thrills of pleasure like those I had when I found my 'first' coyote."

# CHAPTER XVII

# SOME RULES AND THINGS TO REMEMBER

If you are using small animals for bait, use the whole animal, if your method will allow of it, and do not skin the bait, as that will make the coyote or wolf suspicious. Leave the bait, if possible, looking as though it had died a natural death and you will be more successful in your trapping.

---

Do not, if timber wolves are expected, stake a single trap on smooth ground, for the captured animal will be almost certain to escape if you can not visit the trap soon after the animal is caught. This is especially true when using the smaller sizes of traps. When using the regular wolf trap, it may sometimes be fixed solidly if desired but it is better to use a drag of some kind.

---

If you find some animal that the wolves have killed, do not fail to set traps there at once. While it is possible that the wolves will not return, there is a chance, and then one is almost certain to catch coyotes if there are any about.

Wolves are sometimes suspicious of a large bait and will not venture near to it. In such cases one may sometimes make a catch by setting a trap somewhere near by using a small scrap of bait only. The trap may be placed in the open side of a natural half circle of brush, and the bait placed behind it. The tail of a skunk is said to be an unfailing lure in such sets.

---

Sometimes a badger will be caught in a wolf or coyote trap. If so, do not skin it, as they are worth but little; kill it and let it lay on the spot, setting the trap by the side of it. The trap may be set in the loose dirt that the captured badger has dug up and there will be no signs of human interference. It is almost certain that a wolf or coyote will be caught there, within a few nights.

———

When you find where the animals are traveling on trails, if there is not much stock about, to interfere with the traps, make a set on the trail, without bait. Such a set is very good for the old, wary animals.

———

As a general rule, it is best to use blind and scent sets in summer, when the weather is warm and bait soon becomes tainted. The wolves are likely to pass tainted bait by with a sniff, although the coyote is not so particular, and at times prefers carrion. In summer, too, food is more plentiful and the animals are not likely to be hungry. In winter it is best to use bait, as then it will remain fresh for a considerable length of time and the wolves are hungrier at that time.

———

Of meat baits, horse flesh is perhaps the best, and next in order comes antelope. Beef, pork, mutton, and the flesh of all game animals is also good for bait and the young animals are always preferred and selected, if the wolves do the killing. They do not like the flesh of old or diseased animals. Jack rabbits, cotton-tails, prairie dogs, badgers and sage hens make good bait for wolves and of these the jack rabbit is preferred, perhaps because it contains so much blood.

———

It is a good idea to have some small traps, No. 1, with which to catch prairie dogs for bait. The animals are rather wary, however, and care must be used in setting and covering. A 22 caliber rifle is also useful for procuring bait.

When tending the traps, one should carry a long range rifle as he will get shots at coyote, wolf or badger nearly every day. The animals killed in that way add considerable to the income of some of the western wolfers.

---

There will be but little chance of making a catch as long as any human scent or signs remains about the setting. The scent will pass away within a few days, but one should always guard against leaving signs. A rain, or a fresh fall of snow will sometimes help the trapper out, as it removes or covers all signs of human presence. Some broken weeds or a freshly crushed lump of ground will alarm the animal, and through such apparently trifling causes, one may fail to make a catch.

---

When looking at the traps go on horseback and do not dismount unless it is absolutely necessary. On horseback, one may ride up quite close to the trap and the wolves will not be alarmed. If, however, it is necessary to go on foot, do not approach the traps nearer than necessary to see if you have made a catch, also do not go oftener than need be.

---

Sometimes a coyote will uncover a trap, or dig it up from its bed. There is no way to prevent this and the only hope of catching the animal, is in having other different sets in the same locality. Some other method may catch him. For the same reason we would advise the trapper to make use of different sets when putting out the traps, for the method that will catch one would not be successful with another.

---

Do not depend on a few traps alone. Have all that you can look after. If one chance is good, two are better, and those who make the largest catches are the diligent workers, who run long lines.

---

Wolves, like all other wandering animals, have a regular route of travel. While they may vary somewhat from this course, they are sure to continue in the same general line so that when you see tracks in any locality, you may be certain that the animal will travel somewhere near there again.

---

When setting a trap, never leave it until you are satisfied that it is as near a perfect set as can be made. If you do that way, you are sure to be successful.

---

Whenever possible, make the set on the windward side of the wolfs route, that is, on the side from which the prevailing winds blow. In that way the animal is more certain to scent the bait, and will easily follow it up wind to the trap.

---

Some wolfers make it a practice to burn bones and other animal matter near the camp at night, believing that it will draw wolves into the vicinity.

---

All of the foregoing rules will help, and should be kept in mind, but what is more important than any of them is that one be industrious and observing, always endeavoring to learn more of the habits and nature of the animals he seeks for. Such a one is bound to make a success of wolfing.

# CHAPTER XVIII

# THE TREACHEROUS GREY WOLF

### By Perry Davis

The accompanying photo shows the writer holding up the skins of two mighty greys; either wolf would have weighed a hundred pounds, and measured six feet from tip to tip. Little does the average person know of the great damage done by these destructive and blood thirsty desperadoes of the western stock range. Cowardly and evasive, when coming in contact with men, yet when these two blood thirsty companions were running at large, were capable of torturing a full grown cow to death; sometimes a bunch of them will destroy good sized horses. The swift footed and aggressive range steer, equipped with nature's weapons, his long sharp horns, falls an easy victim to the powerful jaws, sharp teeth, and the wise generalship of these terrible brutes.

Five wolves have been killed in this community last winter, and there is but little sign of others, and no complaints from the stockmen. Billy Clanton claims to have lost about 40 head of cattle, mostly calves and yearlings in the last eighteen months and he blames this small bunch of wolves for that loss. The great state of South Dakota pays the miserable sum of $5.00 bounty on grey wolves and $2.00 on coyotes. Last year the bounty claims were paid 80 cents on the dollar, as the claims were in excess of the fund appropriated for bounty purposes.

I have heard of wolves attacking persons in the woods of the Northeastern States; I have no reason to doubt this — they may be a different wolf from our grey wolf, or buffalo wolf, as they are often called. I have seen them in the Panhandle country of Northwest Texas, in Colorado, Wyoming, the Dakotas, Montana and Canada and they are all the same, as far as I could see, in looks, size and habits, and I have never heard of them molesting anyone in the above mentioned places. Of course, there is the coyote, he is everywhere I have ever been and some call him a wolf. Fur dealers call him prairie wolf; frequently some fellow will tell me about a black wolf, or a big white one, but I just let him run it over me; I don't tell him he is a prevaricator, neither do I get angry and try to kill him. I permit him to think he is telling me something and try to look unconcerned and solemn, but I think he has looked down on the back of a grey wolf from high ground and he looked dark and the more he thought about it, the darker it became, until he became almost too black for anything. The same wolf standing on a hill above you, will show the white and yellow on his breast and belly and that always looks so much like that big white wolf. I do not doubt but that there is an occasional black wolf, but I have never seen one.

I want to see every wolf and coyote in the country with his hide nailed up to dry. I did not encourage others to trap when I was wolfing, as I wanted to know how to work my range to the best advantage, and beginners often make them hard to catch; their work is too coarse and the wolves get wise. To the boys who inquired in the July number about methods of setting and baiting for wolves, I will say I will give you the best I've got. While an experienced wolfer can give you some good pointers, he can do you no good, unless you are an early riser and an energetic worker with lots of patience, for successful wolfing is not a lazy man's job. Of course, I do not know anything about trapping in the woods or in the country east of the Missouri. No. 4 Newhouse traps are the best where you are trapping wolves and coyotes both.

A prairie dog town is a good place, especially if the country is rough around it, as wolves come to catch the dogs. Make a blind set on some smooth mound, set about three traps close together. Kinsey stakes all three to one pin, probably to save time, but I always stake them so that they can't quite pull them together but it takes more work. The wind generally blows from the northwest and wolves generally come to a setting facing the wind, and you will see the advantage in having your traps set on the "windward" side or set them in a triangle with bait in center — a prairie dog cut in several pieces and then put together to look natural. In picking the pieces up, he is liable to step around some. If the dog is whole, he may carry it away without being caught. It is not always necessary to bait after you have caught one, as he leaves scent that will attract others. Get traps in bare ground, don't chop out places in the grass. In trapping along trails and creeks always remember the wind; this is important. Roll up a bunch of wool to put under the pan and cover the whole trap with dry dirt, especially in winter.

If you have been covering your traps with paper, cut it out — wool is more convenient and the mice do not uncover your trap and the wind does not uncover it so much. If you are bothered by having cattle spring your traps at a carcass, set your trap under the edge of the carcass where stock will miss them but when the coyote rears back to pull off a bite, it is right where he will put his front feet. I have often killed "Big Jaws," old horses and cripples and then set traps on the trails they follow to feed on the carcass, but seldom set the trap at the carcass. Good strychnine is good if one knows how to use it. If you want to make drop baits, cut up small pieces of the paunch and roll the poison up in it. They like that part of an animal and if they swallow it while it is frozen, it will unroll in the stomach and give the poison a chance to act quickly.

I often use a light wagon in setting traps and sometimes carry dirt to cover with. I throw a wagon sheet out to stand on and do all the work without stepping on the ground, as one should always leave as little scent as possible. I think that most kinds of scent are good or anything that smells rotten enough, but the old grey is certainly cunning and hard to trap, especially if he has lost a few toes. There are grey wolves that do not kill cattle; when I commenced to hunt wolves, I studied them very carefully. I opened and examined the stomach of all I caught and instead of finding them loaded with fresh meat, I found over half without anything in the stomach at all; others had pieces of bones, grass and old pieces of hide stripped from old dry carcasses and I found rabbits, mice and gophers and this was in the lower Musselshell Country where there were thousands of cattle.

I have tried hounds, and have had some of the best that I could get but they were never successful. I never had hounds that would kill a grown wolf, but they often stopped the wolf until I could shoot it and I never knew them to make a good fight more than once, besides dogs knock their toe nails off on rocks and get crippled up with cactus and often a whole pack will almost ruin themselves by killing porcupines, the quills getting in the throat and sometimes will work through the head and into the eyes and blind them. I can take traps and beat any bunch of dogs I ever tried for both wolves and coyotes.

A wolf hound is often very stupid and does some very laughable things. I had six good ones on a trip in Canada. I was going down the Medicine Lodge Valley, had team and the hounds; on each side of the road about three hundred yards ahead were a bunch of cattle, near each bunch there was a coyote. I tried to send the dogs after them but they could not see them, as they were sitting still. Just then the dogs saw a badger about a quarter of a mile down the road, and they were not long getting there. As they passed the cattle, both coyotes started after the dogs and followed them to within a few steps of the scene of battle, where the six dogs were tearing at the tough skin of the badger. The coyotes seemed to think it was "heap fun" and then one coyote jumped into the fight and out again and then the other and they repeated it several times, when at last a young dog discovered one of the coyotes and started him over a hill and the other coyote following at the heels of the dog.

Finally the hound found that he was out-numbered and went back; the other five never knew that there had been a coyote in the valley, but were still tearing away at the dead badger as I drove up. Well, I felt like saying something, but I didn't.

# CHAPTER XIX

# WOLF CATCHING

This article by R. H. Winslow was originally contributed to the HUNTER-TRADER-TRAPPER, but being of special interest is reprinted here:

"It was my misfortune sometime ago to contract a nervous disorder, which quite incapacitated me. After securing the medical advice of one of the world's best specialists, it was apparent that I would find health, if at all, only in a 'journey to nature.' Accordingly I decided to leave New York and spend a year in the West, there to hunt quail, prairie chicken, wild turkey, rabbits, bob cats, wolves, deer and bear.

"At first I went to Oklahoma and from there traveled by easy stages to the Mill Iron Ranch in Northwest Texas, which I have thus far made my headquarters.

"The feathered tribe, rabbits, prairie dogs and bob cats interested me for a while, but soon my thoughts became centered on wolves. Indeed, they are extremely interesting, and I was not long in discovering that it would be necessary to cope with animals of almost human intelligence. Too, they were quite plentiful — could be seen any day while riding over the plains — and night they made hideous with their howls. Would I hunt them with horse an gun, horse an dogs, or attempt to trap them? That was the question confronting me.

"My first experience with horse and gun came about in this way: Two young cowboys, Ernest Edwards and Robert Russell, were with me hunting prairie chicken; we saw a wolf lying in the sage grass about five hundred yards away, and decided that although we had shot guns, we would endeavor to ride up sufficiently close to get a shot. Edwards and I were within about eighty yards of the wolf when he started; both fired, and Russell started immediately in pursuit. Russell ran after him for about three miles, when the chase was taken up by Edwards, who, upon his famous sorrel, 'Playmate,' was soon within a few yards of him and fired with his shot gun. Three shots brought him to the ground.

"After this I saw cowboys try to rope wolves, but seldom with success; and frequently they would attempt to kill them from their mounts with carbine or revolver, but were likewise seldom successful. It was not long, accordingly, before it was evident to me that very little success would attend my efforts with horse and gun.

"The next plan was to try riding to the hounds. There are on the ranch many imported wolf-hounds, two grey hounds and two blood hounds. It comprises about a million acres and these dogs are allowed to roam over it at will; sometimes they are at Estellme; sometimes at Shamrock; sometimes at Aberdeen; sometimes at other places. There is no regular hunting with them by the foremen or cowboys, and none of the owners live on the ranch. These hounds are perfectly trained, though, and understand quite well the ways of a wolf. The following is my first experience with horse and dogs: "The day before my arrival at the Beasley Camp, which included a house of a dozen or so rooms, barns and the like, a beef had been killed and the waste left laying about a hundred yards from the house. We had just gone in to luncheon when one of the boys noticed a large wolf going up to eat upon this waste. Within an incredibly short time we were out of our seats, some yelling for the dogs which were lying around the porch, and others straddling the horses already saddled. The chase was on. It lasted, however, for only about twenty minutes, for the wolf was soon 'picked up.' After this we had several other chases.

"Formerly, hunting with hounds here was practicable and extremely interesting, but now that there are wire fences everywhere it is quite impossible to follow the dogs, and, moreover, when after a wolf they frequently leave the ranch and go upon the premises of some 'nester' (farmer) who has planted poison.

"In a pack of a dozen dogs, say, there are generally two grey hounds used as 'tripping' dogs; that is, they run ahead of the main body and trip or throw the wolf, sometimes twice — so the others have time to come up and jump on. Generally they do no fighting themselves.

"The last plan was to try trapping, and I have found that most successful.

"I found that, first, it was necessary to boil the traps, preferably in blood, so as to kill the odor of steel; secondly, that my gloves and the soles of my boots should be dipped in blood, so as to kill all human scent; thirdly, that I should prepare a large number of round logs, about four feet long and weighing about forty pounds, with a notch in the middle of each, to receive the chain. Then came the consideration of bait.

"At first I used no bait but depended solely upon trail setting and for the following reasons: A trapper who was formerly in the employ of the Hudson Bay Company told me of a setting by which he attained the greatest success, and it is as follows: Take a forked stick the shape of a V, the prongs being about two and a half feet long and with knots or projections on them; fit this V around a mesquit bush so the bush will be pressed closely into the sharp part of the V; place the bait, preferably a rabbit — close against the tree and in the sharp part of the V; then set the trap, completely covered, with springs bent inward, eighteen inches back from the bait and in the V, with the chain covered and fastened to the bush. A wolf will go into a V but will never step over anything two inches high to get bait. I tried this setting but without success. The wolves would go nightly within about ten yards of my traps but no nearer.

"Then I tried staking out a cow's head with the stake driven down so it would not project at all above. But before driving the stake in the ground I had the rings attached to my chains on it and under the head. Around this head I set ten traps in a circle. As before, the wolves would go within about ten yards, but no nearer. I decided, therefore, temporarily, to use no bait, but to try trail setting, for nightly two particular paths were literally covered with wolf tracks.

"My traps, logs, gloves and boots having been prepared, they were taken in a wagon to places for settings; the traps were sunk into the ground so that when leveled there was about a quarter of an inch of dirt on top of the tredles; then the chains were sunk; and finally the logs. About the setting: The center of the tredle should be in the center of the trail; place under the tredle a piece of cotton — over it, a round piece of paper twice its size with a place cut out over the restraining lever; cover very carefully and be quite sure there are no lumps to get caught between the jaws when thrown; and, lastly, leave no loose soil visible so there will be no trace whatever of any disturbance of the earth. Three traps should be set in a row with the jaws, when set, six inches apart. This plan was entirely successful, and I caught wolves nightly. In using a log such as has been described there should always be used with it the two-pronged drag such as is furnished with the No. 4 1/2 Newhouse traps. A wolf may get a few hundred yards away, but he will never break loose, and may be traced quite easily. It is unnecessary but I use a bloodhound on the ranch, 'Red,' for this purpose. With a stationary fastening something may break.

"In time it became my good fortune to drift around to the bull pasture where Curtis Brown, a nice young cowboy, is feeding cotton seed to half a thousand bulls. Here I found trail trapping almost impracticable on account of the bulls following the trails and throwing the traps, and because, seemingly, the wolves would go directly to the carcass of a dead bull without reference to any trail. Accordingly I would watch the carcass closely (about twenty bulls have died) and wherever a wolf had begun to eat on a carcass I would set my traps so as to catch him when he returned to his meal. This plan has been all one could ask.

"Finally, I tried luring wolves to my bait by setting four traps in a row as described in trail setting; but between the second and third I buried a bone or lump of meat which had been allowed to roast and smolder all the night before. Wolves could smell this miles away, would come to it and get in the traps. This, indeed, is the best scheme I know anything about.

"I have noticed that Mr. Ernest Thompson Seton and others say a she wolf or dog staked out in the mating season is an infallible lure; and a captive wolf that will howl is good at any time.' We have a number of female wolves around the camp now and have had them for a long while, one is quite gentle and they howl. They have been staked out frequently with a circle of traps around each, but no wolf has been near.

"Aside from the sport to be obtained in trapping wolves, the pecuniary feature is of interest to the trappers. In New Mexico where they are much more plentiful than in Texas, there is a bounty of twenty dollars each on Lobo wolves (Canislupus) and two and one-half dollars on coyotes. Moreover the trapper does not have to wait for his money for the large ranch owners pay cash for the scalps in order to have him trap on their range, thus decreasing the number of wolves and thereby protecting their cattle and sheep. Too, the trapper is usually furnished a horse or two."

# CHAPTER XX

# WITH THE COYOTES

## By Louis Wessel

While the tourist speeds across the cheerless plains on his way westward, snugly seated in the upholstered berths of an overland limited, the objects of attraction over the landscape are so rare that he will find little desire to spend or waste, as he will say, much time in viewing the scenery; and instead, will settle down to a book or something or other less monotonous than that almost boundless stretch of country, through which he must pass, before he can expect to see the rugged peaks of the Rockies loom up about the distant horizon. Swiftly the limited is carrying him toward his destination, yet slowly very slowly the time passes for him, as hour after hour wears away without bringing a change of scene, until even the monotony of the situation begins to generate in him an interest for the surroundings.

He lends a closer scrutiny to the objects as they speed by. "Why is yonder bluff so lifeless and dreary?" he mused. "What fantastic forms are those near it?" They are but spurs of the famous "Bad Lands." "And this large field of bushes, what is it," he inquires. Some newly formed friend who is better acquainted with the nature of the Great Plains will inform him that this is but a patch of sage bush, an aridity loving plant, characteristic of this region. He will explain that yonder mounds are part of a prairie dog town, and the little marmot like forms, perched in their peculiar attitudes on the little round knolls, represent the inhabitants of this populous city. The traveler has oft heard of prairie dogs, and is surprised on a close acquaintance with them. They appear so different from what his mind has pictured them. He watches them scamper to their burrows, sit up for a moment on their haunches and dive out of sight.

His interest, however, is not completely aroused until he

catches sight of a dog like form, half hidden among the sage bush. He watches it as it disinterestedly trots along with drooping head and tail, a picture of despair, most perfectly suited to its environments. Once it stops all alert, looks back over its shoulder, ears pointed and nose uplifted, and the train leaves it behind in all its loneliness. This is our first acquaintance of the coyote or prairie wolf. Coyotes are of several varieties, each differing from the rest through certain peculiarities in form, size or color, and each having a well defined geographical range. Collectively they range from the upper Mississippi Valley westward through the Great Plains and Rocky Mountains, southward to northern Mexico and northward into British Columbia and the Northwest Territories.

While the coyote is found in one or another of its forms, in greater or lesser numbers throughout this region, its most congenial home is among the Bad Lands and among the sandstone ridges, steep sided buttes and deep narrow coulees and canyons in the Colorado and upper Missouri Valleys, and it is here that its greatest numbers are found. Being thoroughly fitted to these surroundings it has been enabled to hold its own through the advent of civilization, while most of its larger co-inhabitants have been sadly reduced in numbers.

It is true that the combined actions of poisons, traps and high power rifles have done much to reduce the numbers of the coyote in some of its favorite haunts, yet, in other localities, its persistent numbers are deserving of considerable credit. They prove but the survival of the fittest.

Among the mountains the coyote is rarely found, though since the coming of the white man with his flocks they have multiplied considerably in several localities even to such an alarming degree that ranches have found it unprofitable to further attempt to raise sheep.

The coyote of the plains is considerably smaller than the wolf, being intermediate in size between the red fox and the grey wolf. It has the short body, bushy tail, rounded head and pointed nose of the fox and might easily be mistaken for one. Its general color is fulvous, grizzled with black and white hairs and lighter underneath a color remarkable for its ability to blend with the brown and grey, that the arid Plains are clothed in the greater part of the year.

Although well proportioned and being where food is usually plentiful, it rarely fattens up, and almost invariably presents a hungry, half fed appearance. Its food consists mainly of small rodents and birds, such as it can dig up from the ground, or waylay by cat-like maneuvers. Preferring to live on a diet of such animals as it is enabled to capture and kill, it resorts to many schemes and tricks to satisfy its desire for fresh meat. Field mice and gophers living in shallow burrows, fall an easy prey to its diggings. Prairie dogs and cotton tails are waylaid at their place of refuge, and grouse and small birds are pounced upon when they venture too near its place of ambush.

Not always, however, is the coyote enabled to capture its game by such easy means, and when it chooses to dine on jack rabbit, it finds it requires all the power of perseverance and endurance it is capable of mustering up to overtake that fleet creature. As it happens, it is often obliged after a long chase to give up its quarry for a humbler meal. Probably it then decides it is not worth while to hunt the jack alone today, for it knows that if it can persuade one of its comrades to join the chase, Mr. Jack is doomed. When hunting in pairs, they give chase in turns, each stopping to rest in turn, thereby having a double cinch on the poor jack rabbit which is compelled to run continually until exhausted.

In the winter when birds are scarce and the small mammals have hibernated or are huddled away under the snow and frozen ground, the coyote is often sorely pressed for food, and he is then forced to content himself with gnawing off an existence from the frozen carcass of a horse or cow that has died probably months before. His ingenuity of last summer is replaced by a stubborn perseverance, which keeps him traveling day and night in search of scraps of food.

In the spring after the young are born, the bitch is kept busy from morn till night trying to satisfy the hunger of her growling litter of pups, and in her frantic efforts to do so, scruples little on running down and killing a stray sheep or an unprotected calf or colt. When, however, this large prey fails and the smaller game proves insufficient, she is again forced to the humbler larder of some carcass she has discovered on one of her many haunts.

Coyotes are not adepts at burrowing, yet, some credit must be accorded them for work in this line. They often follow up mice and gophers for several feet under the sod, though it remains for the female to exhibit the powers of burrowing possessed by her tribe. In late winter in the southern part of her range, and in the early spring in the northern part, she selects a safe location, usually under a boulder or a ledge of rock, or on the face of a rounded point in a coulee or gulch, from where she may keep a sharp lookout, and sets to work to dig a home for her prospective family. Large quantities of dirt are deposited at the mouth of the burrow, yet this amount is remarkably small when compared with the tunnel from which it is removed, which is often twenty feet or more in length and wide enough to admit a boy, or even in some cases a medium sized man.

At the end of the burrow, which is usually elevated, is an enlargement, in which a litter of from three to eight are brought forth. These are blind and helpless, yet after the first day of their earthly career it seems to become necessary that they exercise both their lungs and limbs, and except for the time that is spent in actual sleep, they keep up a persistent scrambling, one over another, and at the same time a constant growling and whining. The cries of the young and the shuffling about of awkward feet can often be distinctly heard at the mouth of the burrow. This is one of the tests the "wolfer" relies on when he has made the find of a burrow with fresh signs.

As soon as the little ones' eyes are open and their legs grow stronger, they begin to travel, first up and down the burrow, a little further each time, until the mouth is reached. Later on, during the warm sunny days they may be seen playing on the hillside near their home like so many kittens. Before they are half grown the fond mother leads her family out for its initial trip, usually to the nearest watering place, to which they subsequently make regular trips.

It is a pleasing sight to see the young coyotes in playful antics jump up the mother's side and play with her tail as they follow her or chase each other around the bushes. As soon as the young are old enough they are taken out and taught the rules and regulations of the hunt, and long before they are full grown they take an active part in the chase.

In late summer the young leave the maternal home in exchange for an independent life, and it may truly be said that the coyote's childhood day's are over, and it must face the stern realities of life with all its serious consequences. It now prefers to live the life of a hermit, with an occasional short interview with its neighbors.

Contrary to the habits of its cousin and neighbor, the wolf, the coyote is not often seen except singly or in pairs, though it is probable that they are more in the habit of congregating during the night, when the eyes of the hunter and his dogs are closed in sleep, and they are at liberty to roam at will. Their stealthy maneuvers are not apt to disclose their presence, and one usually is not aware of the fact that coyotes are near until he is suddenly reminded of it by one of those unearthly screeching, yelping utterances given vent to by the coyote during the long still night. Immediately the call is taken up by some prowler in a different direction, and in turn is repeated by others further away, until the air fairly resounds with that weird cry. Whether uttered in pleasure or in pain, it is one of nature's most unpleasant calls, and embodies all the hopelessness and despair so apparent on the wide plains of the west.

It is hard to describe the cry of the coyote, though a fair idea may be had by imagining a series of sharp, harsh yelps, terminating into a long drawn painfully entreating howl. Often repeated and echoed by several further away, half a dozen are able to produce enough noise to lead a stranger to believe that he is in the midst of a hundred blood thirsty demons who are proclaiming vengeance on any one that might lack of proper protection.

The coyote is detrimental to but a small degree except to the sheep industry. It is true that coyotes, when hard pressed by hunger, have been known to rob the ranches of its poultry or even to kill a calf or colt, but it is on the defenseless sheep and lambs that they commit their greatest ravages.

In some of the western states, where stock raising is an important industry, large bounties have been paid at different times for the destruction of the coyotes, and these bounties, together with those offered by stock associations and private parties, have induced a number of men and sometimes women, too, to make a business of the extermination of the coyote. Where formerly little time or trouble was spared to destroy these pests, now everybody who has an opportunity eagerly sets traps or poisonous baits for them, shoots at them at long range, runs them down with his bronco to ensnare them in the fatal noose of his lariat, or digs them and their families out of the depths of their underground retreat. The result is obvious. But few localities remain where coyotes hold their own in their original numbers.

The coyote is a wary animal and hard to approach within reasonable pistol shot range, and then only an experienced eye can draw a bead through the gun sights on its dull coat against the usual background of brown or grey. They are fleet foot creatures, and anything short of a greyhound, they are apt to leave behind struggling in the dust. Grey hounds and fox hounds are sometimes employed to run them down, and if one is caught out on the open plain by a pack of these hounds it is quickly dispatched. Frightened almost out of his wits, it repeatedly takes a quick glance back over its shoulder at the furious mob pursuing it, only to find that they are each time a little nearer, until it feels the sharp clasp of the jaws of the leader in deathly embrace. What sport this would be to some of our noblemen across the sea.

Like the red fox, the coyote will sometimes form the friendship of the farmer's dog, and once arrived at a mutual understanding amicable relationship is not easily broken.

As has been said, the coyote is swift afoot, but its wind is easily exhausted, and many a one has fallen a prey, through this lack, to the lariat of the hardy cowboy, who desires nothing more exciting for recreation than a rough and tumble chase through a prairie dog town in pursuit of one of these nimble creatures. Imagine the roughly clad westerner with hair and kerchief flying in the breeze, and the magic noose swinging round and round over his head, whooping at the top of his voice and urging his steed on to its best. Imagine him shooting forth that magic noose and see it settle over the coyote's head. A jerk of the hand tightens the rope, and a turn in the horse's course takes the coyote off his feet and drags him along bouncing from mound to mound into insensibility.

Coyotes cannot be said to possess a vicious nature. Armed with a short club, one may safely enter their burrows, and when trapped the same weapon will complete the work, as they are cowardly and rarely show fight.

Though possessing considerable cunning, coyotes are easier trapped than foxes, though they are slow at taking bait. Large numbers, however, are annually poisoned by placing strychnine in the carcasses of animals that have fallen, through old age or otherwise, of which the pangs of hunger are apt to force coyotes to make a meal. The action of strychnine is exceedingly fast, and it is no unusual occurrence to find a dead coyote a few feet from where it had been enjoying a dinner of poisoned meat.

Of all methods resorted to, however, none is highly responsible for the reduction of the coyote as that of digging up the young (and this often gives up the mother too) from the burrows. By one who is versed in coyote habits, the burrows are easily found, and the work of an hour or two with pick and shovel usually forces them to give up their treasures.

Not always, however, are the results so easily and quickly arrived at. The writer well remembers the first litter of pups he was fortunate enough to capture. After a three days' search among the deep coulees, along the upper Missouri, a den was located. But where? In the crevice of a ledge of sand rock. By placing my ear to the mouth of the burrow, I could hear the pups whining. The burrow was too small to admit me, and as it was too late in the day to commence operations, I plugged up the opening lest the bitch should proceed to transfer her young to some other place of refuge during the night. The greater part of the next day a friend and myself spent in enlarging the burrow with sledge and crowbar, and it was not until late in the afternoon that I was able to crawl in far enough and with the aid of a short stick with some nails drawn through the end, to rake out the six young, one by one.

# CHAPTER XXI

# WOLF TRAPPING AN ART

## By Captain Jack O'Connell

For more than 40 years "Old Hank" Morrison has made his home in the lonely cabin on the shore of a small lake miles from any human habitation, in Alger County. I have often visited this strange old chap, and although the frosts of 70 winters has bent his giant form and silvered his hair, his heart is young. His past life I have never been able to fathom, but to judge from the choice books in several languages in his little cabin, I am led to believe there is a romance in the long, long ago.

The writer slowly recovering from a stroke of paralysis, wishing to get outside the confines of civilization, decided to drop in on "Old Hank" recently. I made a trip despite the deep snow and the protest of my doctor. When I pounded on his door it was rather late at night. "Who in ----," and then pausing in astonishment, threw the door wide open and held out his hand. "Hello Jack," he fairly shouted, shaking my hand in real pump handle fashion, and with all the vigor of his mighty frame. "Blest if I ever expected to see you again! Well! well! well!" He helped me put the horse away in good shape, and then got me a regular "bang up" supper despite the late hour.

Next morning after pancakes and coffee, the very first thing to attract my attention, when I stepped outside was two huge wolf pelts nailed to the side of the shack doing duty as the barn. I became interested at once owing to the unusual size and freshness of both. "Fifty dollars in one night is like finding money, eh," remarked the old man.

I asked him how he managed to catch these cunning animals, knowing that others had met with poor success elsewhere. Says he, "I didn't learn the art of wolf trapping by mail — I have been afflicted with the fad of wolf trapping for 30 years, and in pursuit of them, I have learned a few things not observed by other hunters. I may not know it all but I think I have the only successful trick of trapping these cunning animals and any man who will try my suggestions will meet with good success."

Wolves are very suspicious animals, and have a keen scent for human beings. They will sometimes make a wide detour around a place where I have blazed a tree for the purpose of marking a spot I want to again visit. They are very observing and while the scent of a man's trail through the forest is fresh they will not come within many feet of his path. Hunters find in the school of bitter experience that it is no easy matter to catch them in traps. Old trappers will tell you that it is easier to catch the cutest fox than it is to snare the dullest and most stupid wolf. I have followed the same method all my life — I learned the trick from a half breed trapper in the far Canadian Northwest.

I select an open place in the woods or on the edge of the forest. It is necessary to have a knoll or mound near the center or edge of the clearing on which to place the trap, and in plain view of your bait which you propose to place there for the benefit of Mr. Wolf. A piece of venison or ham is about the best bait to use. I hang this on a sapling or tree and high enough from the ground so the wolf cannot get it by jumping. Make no mistake, mind you, regarding the height from the ground. I put it at least eight feet, for I can tell you a wolf is no slouch when it comes to jumping, especially when the reward is a good chunk of meat, and he happens to be hungry.

"Why not plant a trap under the bait," I suggested, in an effort to appear wise. "Not on your life," says he. "Mr. Wolf is always looking out for just such a joke."

Continuing, he says, "I then cut a stake about six feet long — one with a crotch at one end. I sharpen the other end for the purpose of driving it into the ground. The ring on the end of the chain which is fastened to the trap, I slip over the stake up to the crotch. I then drive the stake into the ground so that no part of it is exposed. I place the trap on the highest part of the knoll and then cover it with leaves. I never take the leaves in my bare hands. I use a piece of bark to carry the leaves in and always from some other place than in the immediate vicinity of the trap, for, mind you, the vagabond is quick to detect if the leaves have been disturbed, and will also scent the presence of man if the leaves have been placed there with his hands.

And remember, it is absolutely necessary that no part of trap, chain or stake be left exposed to view. You see, if you leave the top of the stake sticking out, showing where it was cut off, it is enough to make the vagabond of the woods suspicious that there is a "nigger in the fence" somewhere, and he will lose no time in getting into the next township — instead of attacking the bait.

The bait and trap should be from 30 to 40 feet apart — gauge the distance according to the lay of the ground where the trap is set. When the wolf scents the bait, he will approach it with great caution and endeavor to reach it by jumping. After several unsuccessful attempts to reach it, he will proceed to the highest ground in the immediate vicinity of the bait, where he will set himself upon his haunches and set up a great howl, calling every wolf within the hearing of his voice to the spot.

Your trap, you see, is set upon the highest point of this mound or knoll, and a wolf is almost certain to get into the concealed trap. I sometimes set as many as eight traps on a mound in the vicinity of the bait, and I have caught from two to four wolves in a single night in this way. This was in cases, of course, where a pack arrived before the original finder of the bait was caught. You see if they had found him in a trap when they arrived on the scene, they would not come within yards of the place, but would cut out for tall timber at once, even if they did get a whiff of the bait on the sapling.

Wolves are even more easily caught in the spring of the year than at any other time. This is, of course, after the close of the hunting season. They are hunters themselves and prefer to chase and kill their own game and this accounts for the fact that they will seldom ever touch a deer carcass left in the woods by hunters. When the snow is deep they hunt deer by following their tracks for hours, even days, until they finally get their prey into a place where the animal can't run or defend itself. The feast is then on in short order.

Wolves kill more deer in this country than two legged hunters. If the state is going to do the right thing to protect the deer, just let them put a bounty of $50.00 on the wolves in every county in the Upper Peninsula. Then the woods will be full of men with rifles, and in a year or two there wouldn't be a wolf in Northern Michigan.

If the state did this instead of getting out a lot of swell books on the game laws, we would have the deer with us a few years yet But as it is now, the wolves alone will pick the bones of the last deer in this whole Northern Michigan in less than three years from now. Mark you these words, the state now pays $25.00 for every pelt, but it don't seem to induce hunters and trappers to make a business of wolf trapping. Even with plenty of wolves to catch, following the business for a living is one of extreme hardship, but if they put the bounty in the $50.00 notch, then there would be something doing and the hardship would have no terrors to the men who took up the hunt in earnest."

I spent a week with this interesting man. He has over 300 Newhouse traps of all sizes and quite a pile of mink and skunk skins. He said he never trapped for muskrats as he didn't consider them worth while. His forte being mink, otter, skunk, fox and wild cat, with wolves a side line — although it didn't appear as such to me.

He was greatly interested in my 35 caliber Automatic Winchester rifle and when I fired it a few shots for him, as quick as I could, his eyes stuck out like tea cups. "Say, Hank, you ought to get one." "Not if I know myself, them pop guns is all right for dudes and those fellows with that tired feelin'. Old Betsy is good enough for me." So saying, he took down "Betsy" for my inspection. It was a Sharpe's rifle and a good one, too. It shoots a 45-100 Sharpe's special with a 550 grain ball set trigger, open and peep sights, and weighs 12 pounds. And just to show me how she behaved, he blew a two quart jug off a stump at an estimated distance of 500 yards. "How many deer have you ever killed, Uncle?" I asked. "Well, I can't say, Jack, but give me a dead rest and I can plug a dollar every time at 100 yards." "Well, for heaven's sake, how many have you shot at?"

"Well, I can't tell, Jack, but I must have shot at more than 1,000 of 'em at not over 50 yards."

As a pledge of my friendship, I gave him my Marble pocket axe and knife. It was with a heavy heart that I grasped his honest hand to say good bye — perhaps for the last time on this earth. If so, I sincerely hope to meet him in the "Happy Hunting Grounds" to part no more.

**END OF WOLF AND COYOTE TRAPPING**

www.ingramcontent.com/pod-product-compliance
Lightning Source LLC
Chambersburg PA
CBHW011255040426
42453CB00015B/2421